Cram101 Textbook Outlines to accompany:

Organic Chemistry

L. G. G. Wade, Leroy Wade, 6th Edition

A Cram101 Inc. publication (c) 2010.

With Cram101.com online, you also have access to extensive reference material.

You will nail those essays and papers. Here is an example from a Cram101 Biology text:

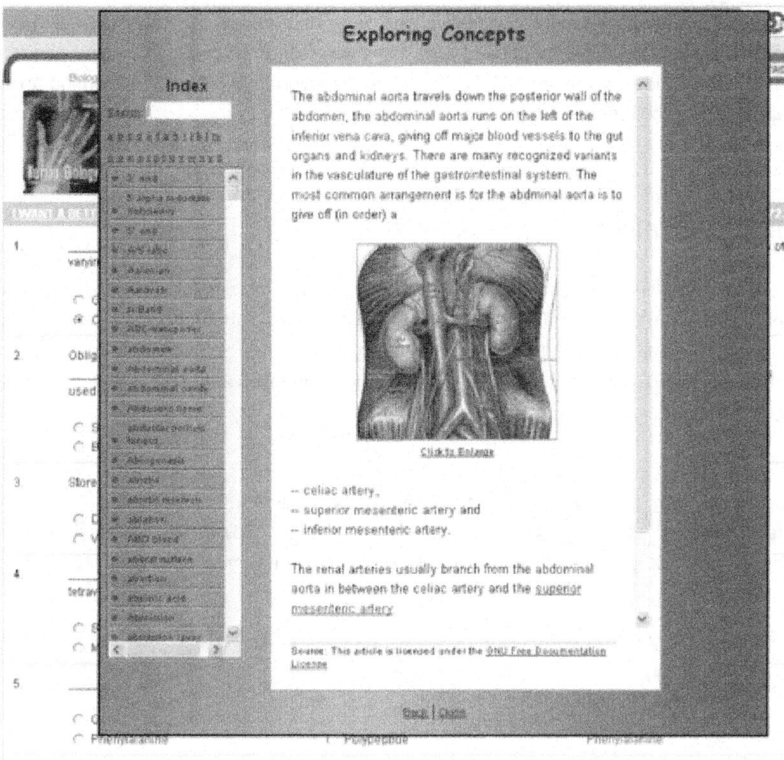

Visit **www.Cram101.com**, click Sign Up at the top of the screen, and enter DK73DW6065 in the promo code box on the registration screen. Access to www.Cram101.com is normally $9.95 per month, but because you have purchased this book, your access fee is only $4.95 per month, cancel at any time. Sign up and stop highlighting textbooks forever.

Learning System

Cram101 Textbook Outlines is a learning system. The notes in this book are the highlights of your textbook, you will never have to highlight a book again.

How to use this book. Take this book to class, it is your notebook for the lecture. The notes and highlights on the left hand side of the pages follow the outline and order of the textbook. All you have to do is follow along while your instructor presents the lecture. Circle the items emphasized in class and add other important information on the right side. With Cram101 Textbook Outlines you'll spend less time writing and more time listening. Learning becomes more efficient.

Cram101.com Online

Increase your studying efficiency by using Cram101.com's practice tests and online reference material. It is the perfect complement to Cram101 Textbook Outlines. Use self-teaching matching tests or simulate in-class testing with comprehensive multiple choice tests, or simply use Cram's true and false tests for quick review. Cram101.com even allows you to enter your in-class notes for an integrated studying format combining the textbook notes with your class notes.

Visit **www.Cram101.com**, click Sign Up at the top of the screen, and enter **DK73DW6065** in the promo code box on the registration screen. Access to www.Cram101.com is normally $9.95 per month, but because you have purchased this book, your access fee is only $4.95 per month. Sign up and stop highlighting textbooks forever.

 ISBN(s): 9781428881716.

Organic Chemistry
L. G. G. Wade, Leroy Wade, 6th

CONTENTS

1. Introduction and Review 2
2. Structure and Properties of Organic Molecules 8
3. Structure and Stereochemistry of Alkanes 28
4. The Study of Chemical Reactions 46
5. Stereochemistry 52
6. Alkyl Halides: Nucleophilic Substitution and Elimination 62
7. Structure and Synthesis of Alkenes 80
8. Reactions of Alkenes 98
9. Alkynes 114
10. Structure and Synthesis of Alcohols 126
11. Reactions of Alcohols 142
12. Infrared Spectroscopy and Mass Spectrometry 158
13. Nuclear Magnetic Resonance Spectroscopy 166
14. Ethers, Epoxides, and Sulfides 172
15. Conjugated Systems, Orbital Symmetry, and Ultraviolet Spectroscopy 190
16. Aromatic Compounds 200
17. Reactions of Aromatic Compounds 216
18. Ketones and Aldehydes 230
19. Amines 240
20. Carboxylic Acids 246
21. Carboxylic Acid Derivatives 260
22. Condensations and Alpha Substitutions of Carbonyl Compounds 272
23. Carbohydrates and Nucleic Acids 284
24. Amino Acids, Peptides, and Proteins 306
25. Lipids 316
26. Synthetic Polymers 322

Organic chemistry

Organic chemistry is a discipline within chemistry which involves the scientific study of the structure, properties, composition, reactions, and preparation (by synthesis or by other means) of chemical compounds that contain carbon. These compounds may contain any number of other elements, including hydrogen, nitrogen, oxygen, the halogens as well as phosphorus, silicon and sulfur.

The original definition of 'organic' chemistry came from the misconception that organic compounds were always related to life processes.

Isomer

In chemistry, Isomer s are compounds with the same molecular formula but different structural formula. Isomer s do not necessarily share similar properties unless they also have the same functional groups. This should not be confused with a nuclear Isomer which involves a nucleus at different states of excitement.

Ascorbic acid

Ascorbic acid is a sugar acid with antioxidant properties. Its appearance is white to light-yellow crystals or powder, and it is water-soluble. One form of Ascorbic acid is commonly known as vitamin C. In 1937 the Nobel Prize for chemistry was awarded to Walter Haworth for his work in determining the structure of Ascorbic acid, and the prize for Physiology or Medicine that year went to Albert Szent-Györgyi for his studies of the biological functions of L-Ascorbic acid.

Morphine

Morphine is a highly potent opiate analgesic psychoactive drug, is the principal active ingredient in Papaver somniferum (opium poppy is considered to be the prototypical opioid. Like other opioids, e.g. oxycodone (OxyContin, Percocet, Percodan), hydromorphone (Dilaudid, Palladone), and diacetylmorphine (Heroin), Morphine acts directly on the central nervous system (CNS) to relieve pain. Morphine has a high potential for addiction; tolerance and both physical and psychological dependence develop rapidly.

Nicotine

Nicotine is an alkaloid found in the nightshade family of plants (Solanaceae) which constitutes approximately 0.6-3.0% of dry weight of tobacco, with biosynthesis taking place in the roots, and accumulating in the leaves. It functions as an antiherbivore chemical with particular specificity to insects; therefore Nicotine was widely used as an insecticide in the past, and currently Nicotine analogs such as imidacloprid continue to be widely used.

In low concentrations , the substance acts as a stimulant in mammals and is one of the main factors responsible for the dependence-forming properties of tobacco smoking.

Carbon

Carbon is the chemical element with symbol C and atomic number 6. As a member of group 14 on the periodic table, it is nonmetallic and tetravalent--making four electrons available to form covalent chemical bonds. There are three naturally occurring isotopes, with ^{12}C and ^{13}C being stable, while ^{14}C is radioactive, decaying with a half-life of about 5730 years.

Configuration

The configuration of a molecule is the permanent geometry that results from the spatial arrangement of its bonds. The ability of the same set of atoms to form two or more molecules with different configurations is stereoisomerism. configuration is distinct from chemical conformation, a shape attainable by bond rotations.

Acetylene

Acetylene is the chemical compound with the formula HC_2H. It is a hydrocarbon and the simplest alkyne. This colourless gas is widely used as a fuel and a chemical building block. It is unstable in pure form and thus is usually handled as a solution.

Dopamine	Dopamine is a neurotransmitter occurring in a wide variety of animals, including both vertebrates and invertebrates. In the brain, this phenethylamine functions as a neurotransmitter, activating the five types of Dopamine receptors -- D_1, D_2, D_3, D_4, and D_5, and their variants. Dopamine is produced in several areas of the brain, including the substantia nigra and the ventral tegmental area.
Ethylene	Ethylene is the chemical compound with the formula C_2H_4. It is the simplest alkene. Because it contains a carbon-carbon double bond, Ethylene is called an unsaturated hydrocarbon or an olefin.
Formaldehyde	Formaldehyde (IUPAC name methanal) is a chemical compound with the formula CH_2O. It is the simplest aldehyde. Formaldehyde also exists as the cyclic trimer trioxane and the polymer para Formaldehyde . It exists in water as the hydrate $H_2C(OH)_2$.
Sugar	Sugar is a class of edible crystalline substances, mainly sucrose, lactose, and fructose. Human taste buds interpret its flavor as sweet. Sugar as a basic food carbohydrate primarily comes from Sugar cane and from Sugar beet, but also appears in fruit, honey, sorghum, Sugar maple (in maple syrup), and in many other sources.
Acetone	Acetone is the organic compound with the formula $OC(CH_3)_2$. This colorless, mobile, flammable liquid is the simplest example of the ketones. Owing to the fact that Acetone is miscible with water, and it serves as an important solvent in its own right, typically the solvent of choice for cleaning purposes in the laboratory.
Acetonitrile	Acetonitrile is the chemical compound with formula CH_3CN. This colourless liquid is the simplest organic nitrile. It is produced mainly as a byproduct of acrylonitrile manufacture. It is widely used as a polar aprotic solvent in synthetic chemistry, and as a medium-polarity solvent in HPLC. Acetonitrile is a by-product from the manufacture of acrylonitrile.
Dimethylacetylene	Dimethylacetylene is an alkyne with chemical formula $CH_3C{\equiv}CCH_3$. Produced artificially, it is a colorless, volatile, pungent liquid at standard temperature and pressure.
Alkene	In organic chemistry, an Alkene olefin, or olefine is an unsaturated chemical compound containing at least one carbon-to-carbon double bond. The simplest acyclic Alkene s, with only one double bond and no other functional groups, form a homologous series of hydrocarbons with the general formula C_nH_{2n}. The simplest Alkene is ethylene (C_2H_4), which has the International Union of Pure and Applied Chemistry (IUPAC) name ethene.
Sodium	Sodium is a metallic element with a symbol Na and atomic number 11. It is a soft, silvery-white, highly reactive metal and is a member of the alkali metals within 'group 1' (formerly known as 'group IA'.) It has only one stable isotope, ^{23}Na.
Sodium acetate	Sodium acetate, (also sodium ethanoate) is the sodium salt of acetic acid. It is an inexpensive chemical produced in industrial quantities for a wide range of uses. Sodium acetate is used in the textile industry to neutralize sulfuric acid waste streams, and as a photoresist while using aniline dyes.

Acetic acid	Acetic acid, CH_3COOH, also known as ethanoic acid, is an organic acid which gives vinegar its sour taste and pungent smell. Pure, water-free Acetic acid is a colourless liquid that absorbs water from the environment (hygroscopy), and freezes at 16.7 °C (62 °F) to a colourless crystalline solid. It is a weak acid, in that it is only partially dissociated acid in aqueous solution.
Nitromethane	Nitromethane is an organic compound with the chemical formula CH_3NO_2. It is the simplest organic nitro compound. It is a slightly viscous, highly polar liquid commonly used as a solvent in a variety of industrial applications such as in extractions, as a reaction medium, and as a cleaning solvent.
Petroleum	Petroleum or crude oil is a naturally occurring, flammable liquid found in rock formations in the Earth consisting of a complex mixture of hydrocarbons of various molecular weights, plus other organic compounds. The term 'Petroleum' was first used in the treatise De Natura Fossilium, published in 1546 by the German mineralogist Georg Bauer, also known as Georgius Agricola. The proportion of hydrocarbons in the mixture is highly variable and ranges from as much as 97% by weight in the lighter oils to as little as 50% in the heavier oils and bitumens.
Alcohol	In chemistry, an Alcohol is any organic compound in which a hydroxyl group (-OH) is bound to a carbon atom of an alkyl or substituted alkyl group. The general formula for a simple acyclic Alcohol is $C_nH_{2n+1}OH$. In common terms, the word Alcohol refers to ethanol, the type of Alcohol found in Alcohol ic beverages. Ethanol is a colorless, volatile liquid with a mild odor which can be obtained by the fermentation of sugars.
Phenol	In organic chemistry, phenols, sometimes called phenolics, are a class of chemical compounds consisting of a hydroxyl group (-OH) bonded directly to an aromatic hydrocarbon group. The simplest of the class is Phenol Phenol - the simplest of the phenols. Although similar to alcohols, phenols have unique properties and are not classified as alcohols (since the hydroxyl group is not bonded to a saturated carbon atom.)
Electrophile	In chemistry, an Electrophile (literally electron-lover) is a reagent attracted to electrons that participates in a chemical reaction by accepting an electron pair in order to bond to a nucleophile. Because Electrophile s accept electrons, they are Lewis acids Most Electrophile s are positively charged, have an atom which carries a partial positive charge, or have an atom which does not have an octet of electrons.
Nucleophile	In chemistry, a Nucleophile is a reagent that forms a chemical bond to its reaction partner (the electrophile) by donating both bonding electrons. Because nucleophiles donate electrons, they are by definition Lewis bases All molecules or ions with a free pair of electrons can act as nucleophiles.

Bond length	In molecular geometry, Bond length or bond distance is the average distance between nuclei of two bonded atoms in a molecule. Bond length is related to bond order, when more electrons participate in bond formation the bond will get shorter. Bond length is also inversely related to bond strength and the bond dissociation energy, as a stronger bond is also a shorter bond.
Linear	In chemistry, the Linear molecular geometry describes the arrangement of three or more atoms placed at an expected bond angle of 180Â°. Linear organic molecules, e.g. acetylene, are often described by invoking sp orbital hybridization for the carbon centers. Many Linear molecules exist, prominent examples include CO_2, HCN, and xenon difluoride.
Dopamine	Dopamine is a neurotransmitter occurring in a wide variety of animals, including both vertebrates and invertebrates. In the brain, this phenethylamine functions as a neurotransmitter, activating the five types of Dopamine receptors -- D_1, D_2, D_3, D_4, and D_5, and their variants. Dopamine is produced in several areas of the brain, including the substantia nigra and the ventral tegmental area.
Ethylene	Ethylene is the chemical compound with the formula C_2H_4. It is the simplest alkene. Because it contains a carbon-carbon double bond, Ethylene is called an unsaturated hydrocarbon or an olefin.
Cycloaddition	A Cycloaddition is a pericyclic chemical reaction, in which 'two or more unsaturated molecules (or parts of the same molecule) combine with the formation of a cyclic adduct in which there is a net reduction of the bond multiplicity.' The resulting reaction is a cyclization reaction. Cycloaddition s are usually described by the backbone size of the participants. This would make the Diels-Alder reaction a [4 + 2 Cycloaddition and the 1,3-dipolar Cycloaddition a [3 + 2 Cycloaddition
Isomer	In chemistry, Isomer s are compounds with the same molecular formula but different structural formula. Isomer s do not necessarily share similar properties unless they also have the same functional groups. This should not be confused with a nuclear Isomer which involves a nucleus at different states of excitement.
2-butene	2-Butene is a compound with formula C_4H_8. It is one of the isomers of butene. It is a good example of cis-trans isomerism.
3-pentanone	3-Pentanone is a colorless liquid ketone with an odor like that of acetone. Its formula is $C_5H_{10}O$. It is soluble in about 25 parts water, and miscible with ethanol and diethyl ether. Two other ketones, which are isomers of 3-Pentanone, are 2-pentanone and methyl isopropyl ketone.
Butane	Butane also called n Butane is the unbranched alkane with four carbon atoms, $CH_3CH_2CH_2CH_3$. Butane is also used as a collective term for n Butane together with its only other isomer, iso Butane (also called methylpropane), $CH(CH_3)_3$. Butane s are highly flammable, colorless, odorless, easily liquefied gases.

Cyclopentane

Cyclopentane is a highly flammable alicyclic hydrocarbon with chemical formula C_5H_{10} and CAS number 287-92-3, consisting of a ring of five carbon atoms each bonded with two hydrogen atoms above and below the plane. It occurs as a colorless liquid with a petrol-like odor. Its melting point is −94 °C and its boiling point is 49 °C. The typical structure of Cyclopentane is the 'envelope' conformation.

Cyclopentane is used in the manufacture of synthetic resins and rubber adhesives and also as a blowing agent in the manufacture of polyurethane insulating foam, as found in may domestic appliances such as refrigerators and freezers, replacing environmentally damaging alternatives such as CFC-11 and HCFC-141b

More advanced technologies, such as computer hard drives and outerspace equipment employ multiply-alkylated Cyclopentane lubricants because of their extremely low volatility.

Enantiomer

In chemistry, an Enantiomer is one of two stereoisomers that are non-superposable complete mirror images of each other, much as one's left and right hands are 'the same' but opposite. Enantiopure compounds refer to a sample having within the limits of detection, molecules of only one chirality.

Enantiomer s have, when present in a symmetric environment, identical chemical and physical properties except for their ability to rotate plane-polarized light (+/-) by equal amounts but in opposite directions.

Chloromethane

Chloromethane R-40 or HCC 40, is a chemical compound of the group of organic compounds called haloalkanes. It was once widely used as a refrigerant. It is a colorless extremely flammable gas with a minorly sweet odor, which is, however, detected at possibly toxic levels.

1-butene

1-Butene is an organic compound and one of the isomers of butene. The formula is C_4H_8.

1-Butene is stable in itself but polymerizes exothermically.

Toluene

Toluene phenylmethane, and Toluol, is a clear water-insoluble liquid with the typical smell of paint thinners, redolent of the sweet smell of the related compound benzene. It is an aromatic hydrocarbon that is widely used as an industrial feedstock and as a solvent. Like other solvents, Toluene is also used as an inhalant drug for its intoxicating properties; however this causes severe neurological harm.

Quinine

Quinine is a natural white crystalline alkaloid having antipyretic , antimalarial, analgesic (painkilling), and anti-inflammatory properties and a bitter taste. It is a stereoisomer of quinidine.

Quinine was the first effective treatment for malaria caused by Plasmodium falciparum, appearing in therapeutics in the 17th century.

Stereocenter

A Stereocenter is any point, though not necessarily an atom, in a molecule bearing groups such that an interchanging of any two groups leads to a stereoisomer. In organic chemistry this term usually refers to a carbon, phosphorus or sulfur atom, though it is also possible for other atoms to be stereocenters in organic and inorganic chemistry.

A molecule can have multiple stereocenters, giving it many stereoisomers.

Alkene	In organic chemistry, an Alkene olefin, or olefine is an unsaturated chemical compound containing at least one carbon-to-carbon double bond. The simplest acyclic Alkene s, with only one double bond and no other functional groups, form a homologous series of hydrocarbons with the general formula C_nH_{2n}. The simplest Alkene is ethylene (C_2H_4), which has the International Union of Pure and Applied Chemistry (IUPAC) name ethene.
Angle strain	The presence of Angle strain in a molecule indicates that in a specific chemical conformation there exist bond angles that deviate from the ideal bond angles required to achieve maximum bond strength. Maximum bond strength results from effective overlap of atomic orbitals in a chemical bond. Angle strain typically affects cyclic molecules because non-cyclic molecules will thermodynamically conform to the most favorable stable state.
Carbon	Carbon is the chemical element with symbol C and atomic number 6. As a member of group 14 on the periodic table, it is nonmetallic and tetravalent--making four electrons available to form covalent chemical bonds. There are three naturally occurring isotopes, with ^{12}C and ^{13}C being stable, while ^{14}C is radioactive, decaying with a half-life of about 5730 years.
Formaldehyde	Formaldehyde (IUPAC name methanal) is a chemical compound with the formula CH_2O. It is the simplest aldehyde. Formaldehyde also exists as the cyclic trimer trioxane and the polymer para Formaldehyde . It exists in water as the hydrate $H_2C(OH)_2$.
Iodide	An Iodide ion is an iodine atom with a −1 charge. Compounds with iodine in formal oxidation state −1 are called Iodide s. This can include ionic compounds such as caesium Iodide or covalent compounds such as phosphorus tri Iodide .
Hydrogen	Hydrogen is the chemical element with atomic number 1. It is represented by the symbol H. At standard temperature and pressure, Hydrogen is a colorless, odorless, nonmetallic, tasteless, highly flammable diatomic gas with the molecular formula H_2. With an atomic weight of 1.007 94 u, Hydrogen is the lightest element.
Dimethyl ether	Dimethyl ether is the organic compound with the formula CH_3OCH_3. The simplest ether, it is a colourless gas that is a useful precursor to other organic compounds and an aerosol propellant. Dimethyl ether is also promising as a clean-burning hydrocarbon fuel.
Ethanol	Ethanol pure alcohol, grain alcohol is a volatile, flammable, colorless liquid. It is a psychoactive drug, best known as the type of alcohol found in alcoholic beverages and in modern thermometers. Ethanol is one of the oldest recreational drugs.
Propylamine	Propylamine, also known as n-Propylamine, is an amine with the chemical formula C_3H_9N. Propylamine is a weak base with its K_b (acid dissociation constant) equaling 4.7×10^{-4}.
Sodium	Sodium is a metallic element with a symbol Na and atomic number 11. It is a soft, silvery-white, highly reactive metal and is a member of the alkali metals within 'group 1' (formerly known as 'group IA'.) It has only one stable isotope, ^{23}Na.

Alcohol

In chemistry, an Alcohol is any organic compound in which a hydroxyl group (-OH) is bound to a carbon atom of an alkyl or substituted alkyl group. The general formula for a simple acyclic Alcohol is $C_nH_{2n+1}OH$. In common terms, the word Alcohol refers to ethanol, the type of Alcohol found in Alcohol ic beverages.

Ethanol is a colorless, volatile liquid with a mild odor which can be obtained by the fermentation of sugars.

Alkanes

Alkanes are chemical compounds that consist only of the elements carbon and hydrogen, wherein these atoms are linked together exclusively by single bonds without any cyclic structure Alkanes belong to a homologous series of organic compounds in which the members differ by a constant relative atomic mass of 14.

Each carbon atom must have 4 bonds, and each hydrogen atom must be joined to a carbon atom

Paraffin

In chemistry, Paraffin is the common name for the alkane hydrocarbons with the general formula C_nH_{2n+2}. Paraffin wax refers to the solids with $20 \leq n \leq 40$.

The simplest Paraffin molecule is that of methane, CH_4, a gas at room temperature.

Wax

Wax refers to beeswax or another substance with similar properties. The traditional meaning, beeswax, refers to a substance secreted by bees and used by them in constructing their honeycombs. The term has come to refer more generally to a class of substances with properties similar to beeswax, in respect of

- plastic (malleable) at normal ambient temperatures
- a melting point above approximately 45 °C (113 °F) (which differentiates waxes from fats and oils)
- a relatively low viscosity when melted (unlike many plastics)
- insoluble in water
- hydrophobic

Waxes may be natural secretions of plants or animals, artificially produced by purification from natural petroleum or completely synthetic. In addition to beeswax, carnauba (a plant epicuticular Wax) and paraffin (a petroleum Wax) are commonly encountered waxes which occur naturally.

Cycloalkanes

Cycloalkanes are types of alkanes which have one or more rings of carbon atoms in the chemical structure of their molecules. Alkanes are types of organic hydrocarbon compounds which have only single chemical bonds in their chemical structure. Cycloalkanes consist of only carbon and hydrogen atoms and are saturated because there are no multiple C-C bonds to hydrogenate

Functional group

In organic chemistry, Functional group s are specific groups of atoms within molecules that are responsible for the characteristic chemical reactions of those molecules. The same Functional group will undergo the same or similar chemical reaction(s) regardless of the size of the molecule it is a part of. However, its relative reactivity can be modified by nearby Functional group s.

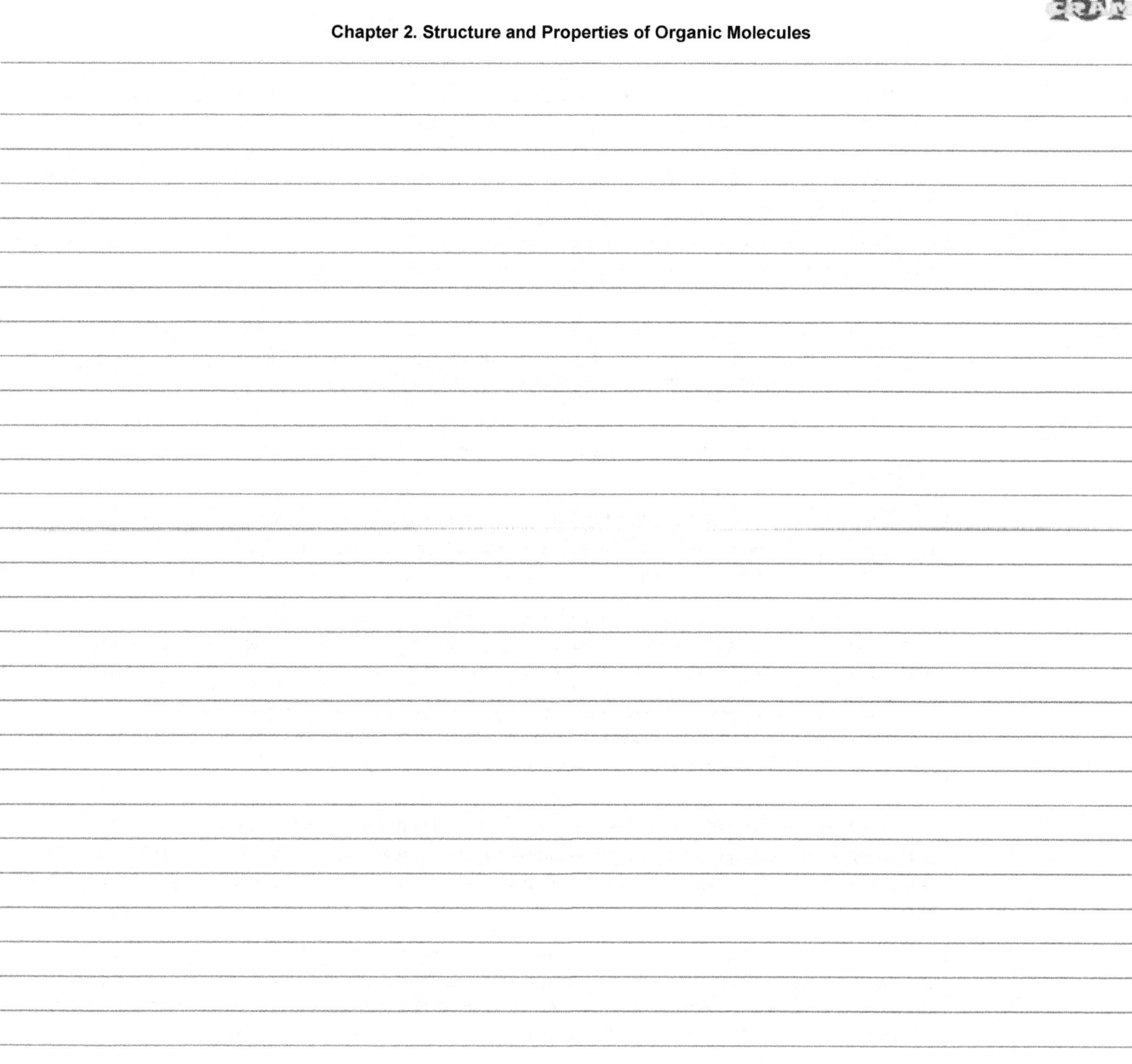

Hydrocarbon	In organic chemistry, a Hydrocarbon is an organic compound consisting entirely of hydrogen and carbon. With relation to chemical terminology, aromatic Hydrocarbon s or arenes, alkanes, alkenes and alkyne-based compounds composed entirely of carbon or hydrogen are referred to as 'pure' Hydrocarbon s, whereas other Hydrocarbon s with bonded compounds or impurities of sulfur or nitrogen, are referred to as 'impure', and remain somewhat erroneously referred to as Hydrocarbon s. Hydrocarbon s are referred to as consisting of a 'backbone' or 'skeleton' composed entirely of carbon and hydrogen and other bonded compounds, and have a functional group that generally facilitates combustion.
Alkyl	In chemistry, an Alkyl group is a hydrocarbon; typically an Alkyl is a part of a larger molecule. The term is usually used loosely, there is no general formula for an Alkyl group. In structural formulae, an Alkyl group is represented with an R. Usually, Alkyl groups resemble hydrocarbons, but with one less hydrogen atom.
Cycloalkene	A Cycloalkene or cycloolefin is a type of alkene hydrocarbon which contains a closed ring of carbon atoms, but has no aromatic character. Some Cycloalkene s, such as cyclobutene and cyclopentene, can be used as monomers to produce polymer chains.
Hexene	Hexene is a higher olefin, or alkene with a formula C_6H_{12}. The 'Hex' is derived from the fact that there are 6 carbon atoms in the molecule, while the 'ene' suffix denotes that two carbon atoms are connected via a double bond. There are several isomers of Hexene, depending on the position of the double bond and the branching of the carbon chain.
Cis-trans isomerism	In organic chemistry, Cis-trans isomerism or geometric isomerism or configuration isomerism or E-Z isomerism is a form of stereoisomerism describing the orientation of functional groups within a molecule. In general, such isomers contain double bonds, which cannot rotate, but they can also arise from ring structures, wherein the rotation of bonds is greatly restricted. The term 'geometric isomerism' is considered an obsolete synonym of 'Cis-trans isomerism' by IUPAC. It is sometimes used as a synonym for general stereoisomerism (e.g., optical isomerism being called geometric isomerism); the correct term for non-optical stereoisomerism is diastereomerism.
Alkynes	Alkynes are hydrocarbons that have a triple bond between two carbon atoms, with the formula C_nH_{2n-2}. Alkynes are traditionally known as acetylenes, although the name acetylene also refers specifically to C_2H_2, known formally (but rarely) as ethyne using IUPAC nomenclature. Like other hydrocarbons, Alkynes are generally hydrophobic but tend to be more reactive.
Benzene	Benzene, or benzol, is an organic chemical compound and a known carcinogen with the molecular formula C_6H_6. It is sometimes abbreviated Ph-H. Benzene is a colorless and highly flammable liquid with a sweet smell and a relatively high melting point. Because it is a known carcinogen, its use as an additive in gasoline is now limited, but it is an important industrial solvent and precursor in the production of drugs, plastics, synthetic rubber, and dyes.
Cyclodecapentaene	Cyclodecapentaene or annulene is an annulene with molecular formula $C_{10}H_{10}$. This organic compound is a conjugated 10 pi electron cyclic system and according to Huckel's rule it should display aromaticity. It is not aromatic, however, because of a combination of steric strain and angular strain.
Cyclohexene	Cyclohexene is a colorless clear liquid cycloalkene with an intense aversive characteristic sharp smell reminiscent of an oil refinery.

It is not very stable upon long term storage with exposure to light and air and should be distilled before use to eliminate peroxides. A common experiment among beginning organic chemistry students is the acid catalyzed dehydration of cyclohexanol with distillative removal of the resulting Cyclohexene from the reaction mixture:

- Refractive Index: 1.4465
- Critical temperature: 287.2 °C (560.4 K)

Cyclopentene

Cyclopentene is a chemical compound with the formula C_5H_8. It is a colorless liquid with a petrol-like odor. It is one of the cycloalkenes.

Ethylbenzene

Ethylbenzene is an organic compound with the formula $C_6H_5CH_2CH_3$. This aromatic hydrocarbon is important in the petrochemical industry as an intermediate in the production of styrene, which in turn is used for making polystyrene, a commonly used plastic material. Although often present in small amounts in crude oil, Ethylbenzene is produced in bulk quantities by combining benzene and ethylene in an acid-catalyzed chemical reaction:

$$C_6H_6 + C_2H_4 \rightarrow C_6H_5CH_2CH_3$$

Approximately 24,700,000 tons were produced in 1999.

Phenyl group

In organic chemistry, the Phenyl group or phenyl ring is the functional group with the formula

C_6H_5-,

where the six carbon atoms are arranged in a cyclic ring structure. This hydrophobic, highly-stable and aromatic hydrocarbon unit can be found in many organic compounds. It can be thought of as being derived from benzene .

Diethyl ether

Diethyl ether is a clear, colorless, and highly flammable liquid with a low boiling point and a characteristic odor. It is the most common member of a class of chemical compounds known generically as ethers. It is an isomer of butanol.

Ether

Ether is a class of organic compounds which contain an Ether group -- an oxygen atom connected to two (substituted) alkyl or aryl groups -- of general formula R-O-R'. A typical example is the solvent and anesthetic diethyl Ether, commonly referred to simply as 'Ether' (ethoxyethane, CH_3-CH_2-O-CH_2-CH_3.)

Ether molecules cannot form hydrogen bonds amongst each other, resulting in a relatively low boiling point compared to that of the analogous alcohols.

Hydroxyl

Hydroxyl in chemistry stands for a molecule consisting of an oxygen atom and a hydrogen atom connected by a covalent bond (single bond.) The neutral form is a Hydroxyl radical and the Hydroxyl anion is called a hydroxide. When the oxygen atom is linked to a larger molecule the Hydroxyl group is a functional group (OH) .

Methanol	Methanol, also known as methyl alcohol, carbinol, wood alcohol, wood naphtha or wood spirits, is a toxic chemical with chemical formula CH_3OH Drinking even small amounts can cause blindness. It is the simplest alcohol, and is a light, volatile, colourless, flammable, toxic liquid with a distinctive odor that is very similar to but slightly sweeter than ethanol (drinking alcohol.)
Acetic acid	Acetic acid, CH_3COOH, also known as ethanoic acid, is an organic acid which gives vinegar its sour taste and pungent smell. Pure, water-free Acetic acid is a colourless liquid that absorbs water from the environment (hygroscopy), and freezes at 16.7 °C (62 °F) to a colourless crystalline solid. It is a weak acid, in that it is only partially dissociated acid in aqueous solution.
Acetone	Acetone is the organic compound with the formula $OC(CH_3)_2$. This colorless, mobile, flammable liquid is the simplest example of the ketones. Owing to the fact that Acetone is miscible with water, and it serves as an important solvent in its own right, typically the solvent of choice for cleaning purposes in the laboratory.
Aldehyde	An Aldehyde is an organic compound containing a terminal carbonyl group. This functional group, which consists of a carbon atom bonded to a hydrogen atom and double-bonded to an oxygen atom (chemical formula O=CH-), is called the Aldehyde group. The Aldehyde group is also called the formyl or methanoyl group.
Butyraldehyde	Butyraldehyde, also known as butanal, is an organic compound with the formula $CH_3(CH_2)_2CHO$. This compound is the aldehyde derivative of butane. It is a colourless flammable liquid with an acrid smell. It is miscible with most organic solvents.
Butyric acid	Butyric acid , also known under the systematic name butanoic acid, is a carboxylic acid with the structural formula $CH_3CH_2CH_2$-COOH. Salts and esters of Butyric acid are known as butyrates or butanoates. Butyric acid is found in rancid butter, parmesan cheese, vomit, and body odor and has an unpleasant smell and acrid taste, with a sweetish aftertaste (similar to ether.) It can be detected by mammals with good scent detection abilities (such as dogs) at 10 ppb, whereas humans can detect it in concentrations above 10 ppm.
Carboxyl group	A Carboxyl group is a set of four atoms bonded together and present in carboxylic acids, including amino acids. Usually abbreviated as either CO_2H or COOH, this set of atoms constitutes a functional group. In every Carboxyl group the carbon atom is attached to an oxygen atom by a double bond and to a hydroxyl group (OH) by a single bond.
Carboxylic acids	Carboxylic acids are organic acids characterized by the presence of a carboxyl group, which has the formula -C(=O)OH, usually written -COOH or $-CO_2H$. Carboxylic acids are Brønsted-Lowry acids -- they are proton donors. Salts and anions of Carboxylic acids are called carboxylates. The simplest series of Carboxylic acids are the alkanoic acids, R-COOH, where R is a hydrogen or an alkyl group.
Cyclohexanone	Cyclohexanone is the organic compound with the formula $(CH_2)_5CO$. The molecule consists of six-carbon cyclic molecule with a ketone functional group. This colorless oil has an odor reminiscent of peardrop sweets as well as acetone. Over time, samples assume a yellow color due to oxidation.
Formic acid	Formic acid is the simplest carboxylic acid. Its formula is HCOOH or CH_2O_2. It is an important intermediate in chemical synthesis and occurs naturally, most notably in the venom of bee and ant stings.

Propionic acid

Propionic acid is a naturally-occurring carboxylic acid with chemical formula CH_3CH_2COOH. It is a clear liquid with a pungent odor. The anion $CH_3CH_2COO^-$ as well as the salts and esters of Propionic acid are known as propionates (or propanoates.)

Propionic acid was first described in 1844 by Johann Gottlieb, who found it among the degradation products of sugar.

Acetamide

Acetamide, CH_3CONH_2, the amide of acetic acid, is a white crystalline solid in pure form. It is produced by dehydrating ammonium acetate:

$$CH_3COONH_4 \rightarrow CH_3CONH_2 + H_2O$$

It is used as a plasticizer and in the synthesis of many other organic compounds.

Acetamide is not extremely combustible, but releases irritating fumes when ignited.

Acetyl

In organic chemistry, Acetyl is a functional group , the acyl with chemical formula $COCH_3$. It is sometimes abbreviated as Ac (not to be confused with the element actinium.) The Acetyl group contains a methyl group single-bonded to a carbonyl.

Acetyl chloride

Acetyl chloride is an acid chloride derived from acetic acid. It has the formula CH_3COCl and it belongs to the class of organic compounds called acyl halides. At room temperature and pressure, it is a clear colorless liquid.

Alkaloids

Alkaloids are naturally occurring chemical compounds containing basic nitrogen atoms. The name derives from the word alkaline and was used to describe any nitrogen-containing base. Alkaloids are produced by a large variety of organisms, including bacteria, fungi, plants, and animals and are part of the group of natural products (also called secondary metabolites.)

Amide

In chemistry, an Amide is one of three kinds of compounds:

- (sometimes called acid Amide the organic functional group characterized by a carbonyl group (C=O) linked to a nitrogen atom (N), or a compound that contains this functional group (pictured to the right); or
- a kind of anion or
- any organic compound derived by the replacement of a hydroxyl group by an amino group.

Amide s are the most stable of all the carbonyl functional groups.

Many chemists make a pronunciation distinction between the two, saying /É™Ë˘miË□d/ for the carbonyl-nitrogen compound and /Ë˘æmaÉªd/ for the anion. Others substitute one of these with , while still others pronounce both , making them homonyms.

In the first sense referred to above, an Amide is an amine where one of the nitrogen substituents is an acyl group; it is generally represented by the formula: $R_1(CO)NR_2R_3$, where either or both R_2 and R_3 may be hydrogen.

Amine

Amine s are organic compounds and functional groups that contain a basic nitrogen atom with a lone pair. Amine s are derivatives of ammonia, wherein one or more hydrogen atoms have been replaced by a substituent such as an alkyl or aryl group. Important Amine s include amino acids, biogenic Amine s, trimethyl Amine , and aniline; see Category: Amine s for a list of Amine s.

Esters

Esters are chemical compounds derived formally from an oxoacid (one containing an oxo group, X=O), and a hydroxyl compound such as an alcohol or phenol. Esters consist of an inorganic acid or organic acid in which at least one -OH (hydroxyl) group is replaced by an -O-alkyl (alkoxy) group. They are analogous to salts, using organic alcohols instead of metallic hydroxides.

Ethyl acetate

Ethyl acetate is the organic compound with the formula $CH_3COOCH_2CH_3$. This colorless liquid has a characteristic sweet smell (similar to pear drops) like certain glues or nail polish removers, in which it is used. Ethyl acetate is the ester of ethanol and acetic acid; it is manufactured on a large scale for use as a solvent.

Methylamine

Methylamine is the organic compound with a formula of CH_3NH_2. This colourless gas is a derivative of ammonia, wherein one H atom is replaced by a methyl group. It is the simplest primary amine.

Nicotine

Nicotine is an alkaloid found in the nightshade family of plants (Solanaceae) which constitutes approximately 0.6-3.0% of dry weight of tobacco, with biosynthesis taking place in the roots, and accumulating in the leaves. It functions as an antiherbivore chemical with particular specificity to insects; therefore Nicotine was widely used as an insecticide in the past, and currently Nicotine analogs such as imidacloprid continue to be widely used.

In low concentrations , the substance acts as a stimulant in mammals and is one of the main factors responsible for the dependence-forming properties of tobacco smoking.

Piperidine

Piperidine is an organic compound with the molecular formula $(CH_2)_5NH$. This heterocyclic amine consists of a six-membered ring containing five methylene units and one nitrogen atom. It is a colorless fuming liquid with an odor described as ammoniacal, pepper-like; the name comes from the genus name Piper, which is the Latin word for pepper. Piperidine is a widely used building block and chemical reagent in the synthesis of organic compounds, including pharmaceuticals.

Triethylamine

Triethylamine is the chemical compound with the formula $N(CH_2CH_3)_3$, commonly abbreviated Et_3N. It is also abbreviated TEA, yet this abbreviation must be used carefully to avoid confusion with triethanolamine, for which TEA is also a common abbreviation. It is commonly encountered in organic synthesis probably because it is the simplest symmetrically trisubstituted amine, i.e. a tertiary amine, that is liquid at room temperature. It possesses a strong fishy odor reminiscent of ammonia.

Acetonitrile

Acetonitrile is the chemical compound with formula CH_3CN. This colourless liquid is the simplest organic nitrile. It is produced mainly as a byproduct of acrylonitrile manufacture. It is widely used as a polar aprotic solvent in synthetic chemistry, and as a medium-polarity solvent in HPLC.

Acetonitrile is a by-product from the manufacture of acrylonitrile.

Benzonitrile

Benzonitrile is the chemical compound with the formula C_6H_5CN, abbreviated PhCN. This aromatic organic compound is colourless, with a sweet almond odour. It is prepared by the dehydration of benzamide, or by the reaction of sodium cyanide with bromobenzene.

	Benzonitrile is a useful solvent and a versatile precursor to many derivatives.
Nitrile	A Nitrile is any organic compound which has a -C≡N functional group. The -C≡N functional group is called a Nitrile group. In the -CN group, the carbon atom and the nitrogen atom are triple bonded together.
Propionitrile	Propionitrile, or ethyl cyanide, is a low-molecular weight nitrile with the molecular formula C_2H_5CN. It is a clear liquid with an ethereal, sweet odor. Propionitrile can be produced by the dehydration of propionamide, by reduction of acrylonitrile, or by distilling ethyl sulfate and potassium cyanide. Propionitrile is poisonous when heated to decomposition, or by treatment with acids.

Alkanes	Alkanes are chemical compounds that consist only of the elements carbon and hydrogen, wherein these atoms are linked together exclusively by single bonds without any cyclic structure Alkanes belong to a homologous series of organic compounds in which the members differ by a constant relative atomic mass of 14. Each carbon atom must have 4 bonds, and each hydrogen atom must be joined to a carbon atom
Alkene	In organic chemistry, an Alkene olefin, or olefine is an unsaturated chemical compound containing at least one carbon-to-carbon double bond. The simplest acyclic Alkene s, with only one double bond and no other functional groups, form a homologous series of hydrocarbons with the general formula C_nH_{2n}. The simplest Alkene is ethylene (C_2H_4), which has the International Union of Pure and Applied Chemistry (IUPAC) name ethene.
Alkynes	Alkynes are hydrocarbons that have a triple bond between two carbon atoms, with the formula C_nH_{2n-2}. Alkynes are traditionally known as acetylenes, although the name acetylene also refers specifically to C_2H_2, known formally (but rarely) as ethyne using IUPAC nomenclature. Like other hydrocarbons, Alkynes are generally hydrophobic but tend to be more reactive.
Cycloalkanes	Cycloalkanes are types of alkanes which have one or more rings of carbon atoms in the chemical structure of their molecules. Alkanes are types of organic hydrocarbon compounds which have only single chemical bonds in their chemical structure. Cycloalkanes consist of only carbon and hydrogen atoms and are saturated because there are no multiple C-C bonds to hydrogenate
Functional group	In organic chemistry, Functional group s are specific groups of atoms within molecules that are responsible for the characteristic chemical reactions of those molecules. The same Functional group will undergo the same or similar chemical reaction(s) regardless of the size of the molecule it is a part of. However, its relative reactivity can be modified by nearby Functional group s.
Hydrocarbon	In organic chemistry, a Hydrocarbon is an organic compound consisting entirely of hydrogen and carbon. With relation to chemical terminology, aromatic Hydrocarbon s or arenes, alkanes, alkenes and alkyne-based compounds composed entirely of carbon or hydrogen are referred to as 'pure' Hydrocarbon s, whereas other Hydrocarbon s with bonded compounds or impurities of sulfur or nitrogen, are referred to as 'impure', and remain somewhat erroneously referred to as Hydrocarbon s. Hydrocarbon s are referred to as consisting of a 'backbone' or 'skeleton' composed entirely of carbon and hydrogen and other bonded compounds, and have a functional group that generally facilitates combustion.
Aromatic Hydrocarbon	An Aromatic hydrocarbon or arene is a hydrocarbon, of which the molecular structure incorporates one or more planar sets of six carbon atoms that are connected by delocalised electrons numbering the same as if they consisted of alternating single and double covalent bonds. The term 'aromatic' was assigned before the physical mechanism determining aromaticity was discovered, and was derived from the fact that many of the compounds have a sweet scent. The configuration of six carbon atoms in aromatic compounds is known as a benzene ring, after the simplest possible such hydrocarbon, benzene.

Homologous series	In chemistry, a Homologous series is a series of organic compounds with a similar general formula, possessing similar chemical properties due to the presence of the same functional group, and shows a gradation in physical properties as a result of increase in molecular size and mass For example, ethane has a higher boiling point than methane since it has more Van der Waals forces(intermolecular forces) with neighbouring molecules. This is due to the increase in the number of atoms making up the molecule.
Butane	Butane also called n Butane is the unbranched alkane with four carbon atoms, $CH_3CH_2CH_2CH_3$. Butane is also used as a collective term for n Butane together with its only other isomer, iso Butane (also called methylpropane), $CH(CH_3)_3$. Butane s are highly flammable, colorless, odorless, easily liquefied gases.
Enantiomer	In chemistry, an Enantiomer is one of two stereoisomers that are non-superposable complete mirror images of each other, much as one's left and right hands are 'the same' but opposite. Enantiopure compounds refer to a sample having within the limits of detection, molecules of only one chirality. Enantiomer s have, when present in a symmetric environment, identical chemical and physical properties except for their ability to rotate plane-polarized light (+/-) by equal amounts but in opposite directions.
Isobutane	Isobutane is an alkane, isomeric with butane. Recent concerns with depletion of the ozone layer by freon gases have led to increased use of Isobutane as a gas for refrigeration systems, especially in domestic refrigerators and freezers, and as a propellant in aerosol sprays. When used as a refrigerant or a propellant, Isobutane is also known as R-600a.
Isopentane	Isopentane, C_5H_{12}, also called methylbutane or 2-methylbutane, is a branched-chain alkane with five carbon atoms. Isopentane is an extremely volatile and extremely flammable liquid at room temperature and pressure. The normal boiling point is just a few degrees above room temperature and Isopentane will readily boil and evaporate away on a warm day.
Neopentane	Neopentane2-dimethylpropane, is a double-branched-chain alkane with five carbon atoms. Neopentane is an extremely flammable gas at room temperature and pressure which can condense into a highly volatile liquid on a cold day, in an ice bath, or when compressed to a higher pressure. IUPAC nomenclature retains the trivial name Neopentane.
Alcohol	In chemistry, an Alcohol is any organic compound in which a hydroxyl group (-OH) is bound to a carbon atom of an alkyl or substituted alkyl group. The general formula for a simple acyclic Alcohol is $C_nH_{2n+1}OH$. In common terms, the word Alcohol refers to ethanol, the type of Alcohol found in Alcohol ic beverages. Ethanol is a colorless, volatile liquid with a mild odor which can be obtained by the fermentation of sugars.
Substituent	In organic chemistry and biochemistry, a Substituent is an atom or group of atoms substituted in place of a hydrogen atom on the parent chain of a hydrocarbon. The terms Substituent, side chain, group, branch, or pendant group are used almost interchangeably to describe branches from a parent structure, though certain distinctions are made in the context of polymer chemistry. In polymers, side chains extend from a backbone structure.

Alkyl	In chemistry, an Alkyl group is a hydrocarbon; typically an Alkyl is a part of a larger molecule. The term is usually used loosely, there is no general formula for an Alkyl group. In structural formulae, an Alkyl group is represented with an R. Usually, Alkyl groups resemble hydrocarbons, but with one less hydrogen atom.
Butyl	In organic chemistry, Butyl is a four-carbon alkyl substituent with chemical formula $-C_4H_9$. It is derived from either of the two isomers of the alkane called butane. Each of the two isomers of butane give rise to two isomers of the Butyl substituent.
Isopropyl	In organic chemistry, Isopropyl is a substituent form of propane, the three-carbon alkyl functional group. As an isomer of propyl, bonding to an R group occurs at the secondary (2°) carbon. Isopropyl is also known as 1-methylethyl (IUPAC); shorthand notations for this alkyl are i-Pr,^{i}Pr, or Pr^{i} (some care has to be taken when using the abbreviation 'Pr', as it is the same as for the chemical element Praseodymium.)
Propane	Propane is a three-carbon alkane, normally a gas, but compressible to a transportable liquid. It is derived from other petroleum products during oil or natural gas processing. It is commonly used as a fuel for engines, oxy-gas torches, barbecues, portable stoves and residential central heating.
Propyl	In organic chemistry, Propyl is a three-carbon alkyl substituent with chemical formula $-C_3H_7$. It is the substituent form of the alkane propane. For example: Propyl ethanoate, also called Propyl acetate. This is Propyl ethanoate, an ester.
Fischer projection	The Fischer projection, devised by Hermann Emil Fischer in 1891, is a two-dimensional representation of a three-dimensional organic molecule by projection. They are used by chemists, particularly in organic chemistry and biochemistry. All bonds are depicted as horizontal or vertical lines.
Kerosene	Kerosene, sometimes spelled kerosine in scientific and industrial usage is a combustible hydrocarbon liquid. The word Kerosene was registered as a trademark by Abraham Gesner in 1854 and for several years only the North American Gas Light Company and the Downer Company were allowed to call their lamp oil Kerosene. It eventually became a genericized trademark.
Natural gas	Natural gas is a gas consisting primarily of methane. It is found associated with fossil fuels, in coal beds, as methane clathrates, and is created by methanogenic organisms in marshes, bogs, and landfills. It is an important fuel source, a major feedstock for fertilizers, and a potent greenhouse gas.
Petroleum	Petroleum or crude oil is a naturally occurring, flammable liquid found in rock formations in the Earth consisting of a complex mixture of hydrocarbons of various molecular weights, plus other organic compounds. The term 'Petroleum' was first used in the treatise De Natura Fossilium, published in 1546 by the German mineralogist Georg Bauer, also known as Georgius Agricola.

The proportion of hydrocarbons in the mixture is highly variable and ranges from as much as 97% by weight in the lighter oils to as little as 50% in the heavier oils and bitumens.

Organic compound

An Organic compound is any member of a large class of chemical compounds whose molecules contain carbon. For historical reasons discussed below, a few types of compounds such as carbonates, simple oxides of carbon and cyanides, as well as the allotropes of carbon, are considered inorganic. The division between 'organic' and 'inorganic' carbon compounds while 'useful in organizing the vast subject of chemistry...is somewhat arbitrary'.

Cracking

In petroleum geology and chemistry, Cracking is the process whereby complex organic molecules such as kerogens or heavy hydrocarbons are broken down into simpler molecules (e.g. light hydrocarbons) by the breaking of carbon-carbon bonds in the precursors. The rate of Cracking and the end products are strongly dependent on the temperature and presence of any catalysts. Cracking, also referred to as pyrolysis, is the breakdown of a large alkane into smaller, more useful alkanes and an alkene.

Hydrogen

Hydrogen is the chemical element with atomic number 1. It is represented by the symbol H. At standard temperature and pressure, Hydrogen is a colorless, odorless, nonmetallic, tasteless, highly flammable diatomic gas with the molecular formula H_2. With an atomic weight of 1.007 94 u, Hydrogen is the lightest element.

Paraffin

In chemistry, Paraffin is the common name for the alkane hydrocarbons with the general formula C_nH_{2n+2}. Paraffin wax refers to the solids with $20 \leq n \leq 40$.

The simplest Paraffin molecule is that of methane, CH_4, a gas at room temperature.

Source

Source Holdings Ltd. is a specialist European-based provider of Exchange Traded Products (ETPs), including Exchange Traded Funds (ETFs) and Exchange Traded Commodities (ETCs.) The company was founded in 2008 by Bank of America Merrill Lynch, Goldman Sachs and Morgan Stanley and is headquartered inside the City of London, on 88 Wood Street.

Source launched its first product offering on the 20th April 2009, although the 3 founders are reported to have started working on the project in early 2008 .

Wax

Wax refers to beeswax or another substance with similar properties. The traditional meaning, beeswax, refers to a substance secreted by bees and used by them in constructing their honeycombs. The term has come to refer more generally to a class of substances with properties similar to beeswax, in respect of

- plastic (malleable) at normal ambient temperatures
- a melting point above approximately 45 °C (113 °F) (which differentiates waxes from fats and oils)
- a relatively low viscosity when melted (unlike many plastics)
- insoluble in water
- hydrophobic

	Waxes may be natural secretions of plants or animals, artificially produced by purification from natural petroleum or completely synthetic. In addition to beeswax, carnauba (a plant epicuticular Wax) and paraffin (a petroleum Wax) are commonly encountered waxes which occur naturally.
Ozonolysis	Ozonolysis is the cleavage of an alkene or alkyne with ozone to form organic compounds in which the multiple carbon-carbon bond has been replaced by a double bond to oxygen. The outcome of the reaction depends on the type of multiple bond being oxidized and the workup conditions. Alkenes can be oxidized with ozone to form alcohols, aldehydes or ketones, or carboxylic acids.
Halogenation	Halogenation is a chemical reaction that incorporates a halogen atom into a molecule. More specific descriptions exist that specify the type of halogen: fluorination, chlorination, bromination, and iodination. In a Markovnikov addition reaction, a halogen like bromine is reacted with an alkene which causes the π-bond to break forming an haloalkane.
Sorbitol	Sorbitol is a sugar alcohol that the body metabolises slowly. It is obtained by reduction of glucose changing the aldehyde group to an additional hydroxyl group. Sorbitol is a sugar substitute often used in diet foods and sugar-free chewing gum, mints and cough syrups.
Ethylene	Ethylene is the chemical compound with the formula C_2H_4. It is the simplest alkene. Because it contains a carbon-carbon double bond, Ethylene is called an unsaturated hydrocarbon or an olefin.
Newman projection	A Newman projection, useful in alkane stereochemistry, visualizes chemical conformations of a carbon-carbon chemical bond from front to back, with the front carbon represented by a dot and the back carbon as a circle The front carbon atom is called proximal, while the back atom is called distal. This type of representation is useful for assessing the torsional angle between bonds.
Configuration	The configuration of a molecule is the permanent geometry that results from the spatial arrangement of its bonds. The ability of the same set of atoms to form two or more molecules with different configurations is stereoisomerism. configuration is distinct from chemical conformation, a shape attainable by bond rotations.
Eclipsed conformation	In chemistry an Eclipsed conformation is a chemical conformation that exists in any open chain single chemical bond connecting two sp^3 hybridised atoms as a conformational energy maximum. This maximum is often explained by steric hindrance, but its origins sometimes actually lie in hyperconjugation (as when the eclipsing interaction is of two hydrogen atoms.) Staggered conformation image right in Newman projection Eclipsed conformation In the example of ethane in Newman projection it shows that rotation around the carbon-carbon bond is not entirely free but that an energy barrier exists.

Staggered conformation

A Staggered conformation is a chemical conformation that exists in any open chain single chemical bond connecting two sp_3 hybridised atoms as a conformational energy minimum. For some molecules such as those of n-butane, there can special versions of staggered conformations called gauche and anti; see first Newman projection diagram in Conformational isomerism. Staggered conformation image right in Newman projection Eclipsed conformation

Strain

In chemistry a molecule experiences Strain when in a chemical conformation there exist unfavorable bond angles or bond distances. Strain energy is released when the molecule can relax to a conformation with less Strain or when the molecule interacts in a suitable chemical reaction.

Several types of Strain exist:

- Angle Strain
- Torsional Strain
- van der Waals Strain

.

Cyclic compound

In organic chemistry, a Cyclic compound is a compound in which a series of carbon atoms are connected to form a loop or ring. Benzene is a well known example. The term 'polycyclic' is used when more than one ring is formed in a single molecule for instance in naphthalene, and the term macrocycle is used for a ring containing more than a dozen atoms.

Cyclopropane

Cyclopropane is a cycloalkane molecule with the molecular formula C_3H_6, consisting of three carbon atoms linked to each other to form a ring, with each carbon atom bearing two hydrogen atoms.

The bonds between the carbon atoms are considerably weaker than in a typical carbon-carbon bond, yielding reactivity similar to or greater than alkenes. Baeyer strain theory explains why: the angle strain from the 60° angle between the carbon atoms (less than the normal angle of 109.5° for bonds between atoms with sp^3 hybridised orbitals) reduces the compound's carbon-carbon bond energy, making it more reactive than other cycloalkanes such as cyclohexane and cyclopentane.

Cis-trans isomerism

In organic chemistry, Cis-trans isomerism or geometric isomerism or configuration isomerism or E-Z isomerism is a form of stereoisomerism describing the orientation of functional groups within a molecule. In general, such isomers contain double bonds, which cannot rotate, but they can also arise from ring structures, wherein the rotation of bonds is greatly restricted.

The term 'geometric isomerism' is considered an obsolete synonym of 'Cis-trans isomerism' by IUPAC. It is sometimes used as a synonym for general stereoisomerism (e.g., optical isomerism being called geometric isomerism); the correct term for non-optical stereoisomerism is diastereomerism.

Diastereomers

Diastereomers are stereoisomers that are not enantiomers (non-superimposable mirror images of each other.) Diastereomers can have different physical properties and different reactivity. In another definition Diastereomers are pairs of isomers that have opposite configurations at one or more of the chiral centers but are not mirror images of each other .

Ring strain

Ring strain is an organic chemistry term that describes the destabilization of a cyclic molecule--such as a cycloalkane--due to the non-favorable high energy spatial orientations of its atoms. Non-cyclic molecules do not exhibit Ring strain because their terminal (end) atoms are not connected to force a particular type of spatial orientation.

Ring strain results from a combination of angle strain, conformational strain or Pitzer strain, and transannular strain or van der Waals strain.

Angle strain

The presence of Angle strain in a molecule indicates that in a specific chemical conformation there exist bond angles that deviate from the ideal bond angles required to achieve maximum bond strength. Maximum bond strength results from effective overlap of atomic orbitals in a chemical bond.

Angle strain typically affects cyclic molecules because non-cyclic molecules will thermodynamically conform to the most favorable stable state.

Cyclobutane

Cyclobutane C_4H_8, with a molecular mass of 56.107g/mol, is a four carbon alkane in which all the carbon atoms are arranged cyclically, hence Cyclobutane Cyclobutane is a gas and commercially available as a liquefied gas.

Cyclobutane s are Cyclobutane derivatives.

Ether

Ether is a class of organic compounds which contain an Ether group -- an oxygen atom connected to two (substituted) alkyl or aryl groups -- of general formula R-O-R'. A typical example is the solvent and anesthetic diethyl Ether, commonly referred to simply as 'Ether' (ethoxyethane, CH_3-CH_2-O-CH_2-CH_3.)

Ether molecules cannot form hydrogen bonds amongst each other, resulting in a relatively low boiling point compared to that of the analogous alcohols.

Cyclohexane

Cyclohexane is a cycloalkane with the molecular formula C_6H_{12}. Cyclohexane is used as a nonpolar solvent for the chemical industry, and also as a raw material for the industrial production of adipic acid and caprolactam, both of which are intermediates used in the production of nylon. On an industrial scale, Cyclohexane is produced by reacting benzene with hydrogen.

Cyclopentane

Cyclopentane is a highly flammable alicyclic hydrocarbon with chemical formula C_5H_{10} and CAS number 287-92-3, consisting of a ring of five carbon atoms each bonded with two hydrogen atoms above and below the plane. It occurs as a colorless liquid with a petrol-like odor. Its melting point is −94 °C and its boiling point is 49 °C. The typical structure of Cyclopentane is the 'envelope' conformation.

Cyclopentane is used in the manufacture of synthetic resins and rubber adhesives and also as a blowing agent in the manufacture of polyurethane insulating foam, as found in may domestic appliances such as refrigerators and freezers, replacing environmentally damaging alternatives such as CFC-11 and HCFC-141b

More advanced technologies, such as computer hard drives and outerspace equipment employ multiply-alkylated Cyclopentane lubricants because of their extremely low volatility.

Cyclohexane conformation	Cyclohexane conformation is a much studied topic in organic chemistry because of the complex interrelationship between the different conformers of cyclohexane and its derivatives. Different conformers may have differing properties, including stability and chemical reactivity. The very first suggestion that cyclohexane may not be a flat molecule goes back a surprisingly long time.
Steroid	A Steroid is a terpenoid lipid characterized by a carbon skeleton with four fused rings, generally arranged in a 6-6-6-5 fashion. Steroids vary by the functional groups attached to these rings and the oxidation state of the rings. Hundreds of distinct steroids are found in plants, animals, and fungi.
Nicotinamide	Nicotinamide, also known as niacinamide and nicotinic acid amide, is the amide of nicotinic acid (vitamin B_3.) Nicotinamide is a water-soluble vitamin and is part of the vitamin B group. Nicotinic acid, also known as niacin, is converted to niacinamide in vivo, and though the two are identical in their vitamin functions, niacinamide does not have the same pharmacologic and toxic effects of niacin, which occur incidental to niacin's conversion.
Nicotinamide adenine dinucleotide	Nicotinamide adenine dinucleotide, abbreviated Nicotinamide adenine dinucleotide$^+$, is a coenzyme found in all living cells. The compound is a dinucleotide, since it consists of two nucleotides joined through their phosphate groups: with one nucleotide containing an adenine base, and the other containing nicotinamide. In metabolism, Nicotinamide adenine dinucleotide$^+$ is involved in redox reactions, carrying electrons from one reaction to another.
Bicyclic molecule	A Bicyclic molecule is a molecule that features two fused rings. Bicyclic molecule s occur in widely in organic and inorganic compounds. Fusion of the rings can occur in three ways: 1. Across a bond between two atoms - for example, decalin (also known as bicyclo[4.4.0]decane), has a C-C bond shared between two cyclohexane rings; 2. Across a sequence of atoms (bridgehead) - for example, norbornane (also known as bicyclo[2.2.1]heptane), can be viewed as a pair of cyclopentane rings that share three of the five carbon atoms; or 3. At a single atom (spirocyclic, forming a spiro compound) Singly fused rings are the most common, and spiro rings are the least common. A bridge is an unbranched chain of atoms or an atom or a covalent bond connecting two bridgeheads in a polycyclic compound.
Carbon	Carbon is the chemical element with symbol C and atomic number 6. As a member of group 14 on the periodic table, it is nonmetallic and tetravalent--making four electrons available to form covalent chemical bonds. There are three naturally occurring isotopes, with ^{12}C and ^{13}C being stable, while ^{14}C is radioactive, decaying with a half-life of about 5730 years.

Cocaine	The first synthesis and elucidation of the cocaine molecule was by Richard Willstätter in 1898. Willstätter's synthesis derived cocaine from tropinone. Since then, Robert Robinson and Edward Leete have made significant contributions to the mechanism of the synthesis.
Decahydronaphthalene	Decahydronaphthalene, a bicyclic organic compound, is an industrial solvent. A colorless liquid with an aromatic odor, it is used as a solvent for many resins. It is the saturated analog of naphthalene and can be prepared from it by hydrogenation in a fused state in the presence of a catalyst.

Cycloalkanes	Cycloalkanes are types of alkanes which have one or more rings of carbon atoms in the chemical structure of their molecules. Alkanes are types of organic hydrocarbon compounds which have only single chemical bonds in their chemical structure. Cycloalkanes consist of only carbon and hydrogen atoms and are saturated because there are no multiple C-C bonds to hydrogenate
McLafferty rearrangement	The McLafferty rearrangement is a reaction observed in mass spectrometry. It is sometimes found that a molecule containing a keto-group undergoes β-cleavage, with the gain of the γ-hydrogen atom. This rearrangement may take place by a radical or ionic mechanism.
Aldehyde	An Aldehyde is an organic compound containing a terminal carbonyl group. This functional group, which consists of a carbon atom bonded to a hydrogen atom and double-bonded to an oxygen atom (chemical formula O=CH-), is called the Aldehyde group. The Aldehyde group is also called the formyl or methanoyl group.
Halogenation	Halogenation is a chemical reaction that incorporates a halogen atom into a molecule. More specific descriptions exist that specify the type of halogen: fluorination, chlorination, bromination, and iodination. In a Markovnikov addition reaction, a halogen like bromine is reacted with an alkene which causes the π-bond to break forming an haloalkane.
Chloromethane	Chloromethane R-40 or HCC 40, is a chemical compound of the group of organic compounds called haloalkanes. It was once widely used as a refrigerant. It is a colorless extremely flammable gas with a minorly sweet odor, which is, however, detected at possibly toxic levels.
Fentanyl	Fentanyl -- brand names include Actiq, Duragesic, Fentora, and Sublimaze -- is a synthetic primary μ-opioid agonist commonly used to treat post-operative and chronic breakthrough pain. It is approximately 100 times more potent than morphine, with 100 micrograms of Fentanyl equivalent to 10 mg of morphine and 75 mg of pethidine in analgesic activity. It has an LD_{50} of 3.1 milligrams per kilogram in rats, 0.03 milligrams per kilogram in monkeys, and an undetermined milligrams per kilogram in humans.
Radicals	In chemistry, Radicals are atoms, molecules, or ions with unpaired electrons on an otherwise open shell configuration. These unpaired electrons are usually highly reactive, so Radicals are likely to take part in chemical reactions. Radicals play an important role in combustion, atmospheric chemistry, polymerization, plasma chemistry, biochemistry, and many other chemical processes, including human physiology.
Reactive oxygen species	Reactive oxygen species are ions or very small molecules that include oxygen ions, free radicals, and peroxides, both inorganic and organic. They are highly reactive due to the presence of unpaired valence shell electrons. Reactive oxygen species form as a natural byproduct of the normal metabolism of oxygen and have important roles in cell signaling.
Alkynes	Alkynes are hydrocarbons that have a triple bond between two carbon atoms, with the formula C_nH_{2n-2}. Alkynes are traditionally known as acetylenes, although the name acetylene also refers specifically to C_2H_2, known formally (but rarely) as ethyne using IUPAC nomenclature. Like other hydrocarbons, Alkynes are generally hydrophobic but tend to be more reactive.

Carbanion	A Carbanion is an anion in which carbon has an unshared pair of electrons and bears a negative charge usually with three substituents for a total of eight valence electrons . The Carbanion exists in a trigonal pyramidal geometry. Formally a Carbanion is the conjugate base of a carbon acid.
Free radical reaction	A Free radical reaction is any chemical reaction involving free radicals. This reaction type is abundant in organic reactions. Two pioneering studies into Free radical reaction s have been the discovery of the triphenylmethyl radical by Moses Gomberg (1900) and the lead-mirror experiment described by Friedrich Paneth in 1927.
Hydroxyl	Hydroxyl in chemistry stands for a molecule consisting of an oxygen atom and a hydrogen atom connected by a covalent bond (single bond.) The neutral form is a Hydroxyl radical and the Hydroxyl anion is called a hydroxide. When the oxygen atom is linked to a larger molecule the Hydroxyl group is a functional group (OH) .
Hydroxyl radical	The Hydroxyl radical OHÂ·, is the neutral form of the hydroxide ion (OH^-.) Hydroxyl radical s are highly reactive and consequently short-lived; however, they form an important part of radical chemistry. Most notably Hydroxyl radical s are produced from the decomposition of hydro-peroxides (ROOH) or, in atmospheric chemistry, by the reaction of excited atomic oxygen with water.
Enol	Enol s are alkenes with a hydroxyl group affixed to one of the carbon atoms composing the double bond. Enol s and carbonyl compounds are in fact isomers; this is called keto Enol tautomerism: The Enol form is shown above. It is usually unstable, does not survive long, and changes into the keto form shown on the right.
Enzymes	Enzymes are biomolecules that catalyze (i.e., increase the rates of) chemical reactions. Nearly all known Enzymes are proteins. However, certain RNA molecules can be effective biocatalysts too.
Hydrogen	Hydrogen is the chemical element with atomic number 1. It is represented by the symbol H. At standard temperature and pressure, Hydrogen is a colorless, odorless, nonmetallic, tasteless, highly flammable diatomic gas with the molecular formula H_2. With an atomic weight of 1.007 94 u, Hydrogen is the lightest element.
Propane	Propane is a three-carbon alkane, normally a gas, but compressible to a transportable liquid. It is derived from other petroleum products during oil or natural gas processing. It is commonly used as a fuel for engines, oxy-gas torches, barbecues, portable stoves and residential central heating.
Ascorbic acid	Ascorbic acid is a sugar acid with antioxidant properties. Its appearance is white to light-yellow crystals or powder, and it is water-soluble. One form of Ascorbic acid is commonly known as vitamin C. In 1937 the Nobel Prize for chemistry was awarded to Walter Haworth for his work in determining the structure of Ascorbic acid, and the prize for Physiology or Medicine that year went to Albert Szent-Györgyi for his studies of the biological functions of L-Ascorbic acid.
Butylated hydroxyanisole	Butylated hydroxyanisole is an antioxidant consisting of a mixture of two isomeric organic compounds, 2-tert-butyl-4-hydroxyanisole and 3-tert-butyl-4-hydroxyanisole. It is prepared from 4-methoxyphenol and isobutylene. It is a waxy solid used in certain amounts as a food additive with the E number E320.

Carbene	In chemistry, a Carbene is a organic molecule containing a carbon atom with six valence electrons and having the general formula RR'C:. Carbene s are classified into two varieties, singlets and triplets. Most Carbene s are very short lived, although persistent Carbene s are known.
Carbocation	A Carbocation is an ion with a positively-charged carbon atom. The charged carbon atom in a Carbocation is a 'sextet', i.e. it has only six electrons in its outer valence shell instead of the eight valence electrons that ensures maximum stability . Therefore Carbocation s are often reactive, seeking to fill the octet of valence electrons as well as regain a neutral charge.
Hyperconjugation	Hyperconjugation in organic chemistry is the stabilizing interaction that results from the interaction of the electrons in a sigma bond (usually C-H or C-C) with an adjacent empty (or partially filled) non-bonding p-orbital or antibonding π orbital or filled π orbital to give an extended molecular orbital that increases the stability of the system. Only electrons in bonds that are β to the positively charged carbon can stabilize a carbocation by Hyperconjugation. The term was introduced in 1939 by Robert S. Mulliken in the course of his work on UV spectroscopy of conjugated molecules.
Inductive effect	The Inductive effect in chemistry is an experimentally observable effect of the transmission of charge through a chain of atoms in a molecule by electrostatic induction. The net polar effect exerted by a substituent is a combination of this Inductive effect and the mesomeric effect. The electron cloud in a σ-bond between two unlike atoms is not uniform and is slightly displaced towards the more electronegative of the two atoms.
Diazomethane	Diazomethane is the chemical compound CH_2N_2. It is one of the more common diazo compounds. In the pure form at room temperature, it is a famously explosive yellow gas, but it is almost universally used as a solution in diethyl ether.
Methylene	Methylene is a chemical species in which a carbon atom is bonded to two hydrogen atoms. Three different possibilities present themselves: • the $-CH_2-$ group, e.g. dichloromethane (also known as Methylene chloride) • the $=CH_2$ group, e.g. methylenecyclopropene, • and the $:CH_2$ molecule, a carbene known as Methylene. A carbene is a highly reactive organic molecule having a divalent carbon atom with six valence electrons. Methylene groups in a chain or ring contribute to its size and lipophilicity.

Enantiomer	In chemistry, an Enantiomer is one of two stereoisomers that are non-superposable complete mirror images of each other, much as one's left and right hands are 'the same' but opposite. Enantiopure compounds refer to a sample having within the limits of detection, molecules of only one chirality. Enantiomer s have, when present in a symmetric environment, identical chemical and physical properties except for their ability to rotate plane-polarized light (+/-) by equal amounts but in opposite directions.
Isomer	In chemistry, Isomer s are compounds with the same molecular formula but different structural formula. Isomer s do not necessarily share similar properties unless they also have the same functional groups. This should not be confused with a nuclear Isomer which involves a nucleus at different states of excitement.
Stereocenter	A Stereocenter is any point, though not necessarily an atom, in a molecule bearing groups such that an interchanging of any two groups leads to a stereoisomer. In organic chemistry this term usually refers to a carbon, phosphorus or sulfur atom, though it is also possible for other atoms to be stereocenters in organic and inorganic chemistry. A molecule can have multiple stereocenters, giving it many stereoisomers.
Absolute configuration	An Absolute configuration in stereochemistry is the spatial arrangement of the atoms of a chiral molecular entity (or group) and its stereochemical description e.g. R or S. Absolute configuration s for chiral molecules are traditionally obtained by X-ray crystallography but only when the compound crystallises in one of the 65 Sohncke Groups (Chiral Space Groups.) Alternative techniques are Optical rotatory dispersion, vibrational circular dichroism and the use of chiral shift reagents in proton NMR. In step two the assignment of R or S is based on the Cahn-Ingold-Prelog priority rules. Absolute configuration s are also relevant to characterization of crystals.
Carbon	Carbon is the chemical element with symbol C and atomic number 6. As a member of group 14 on the periodic table, it is nonmetallic and tetravalent--making four electrons available to form covalent chemical bonds. There are three naturally occurring isotopes, with ^{12}C and ^{13}C being stable, while ^{14}C is radioactive, decaying with a half-life of about 5730 years.
Chirality	In general, chiral molecules have point Chirality at a single stereogenic atom, usually carbon, which has four different substituents. The two enantiomers of such compounds are said to have different absolute configurations at this center. This center is thus stereogenic (i.e., a grouping within a molecular entity that may be considered a focus of stereoisomerism), and is exemplified by the α-carbon of amino acids.
Dichlorocarbene	Dichlorocarbene is a carbene commonly encountered in organic chemistry. This reactive intermediate with chemical formula CCl_2 is easily available by reaction of chloroform and a base such as potassium t-butoxide or sodium hydroxide dissolved in water. A phase transfer catalyst for instance benzyltriethylammonium bromide is added to facilitate the migration of the hydroxide in the organic phase.

1-bromobutane

1-Bromobutane$_2$CH$_2$Br) is a colorless liquid that is insoluble in water, but soluble in ethanol and diethyl ether. As a primary alkyl halide, it is especially prone to S_N2 type reactions. It is commonly used as an alkylating agent, or in combination with magnesium metal in dry ether (Grignard reagent) to form carbon-carbon bonds.

2-bromobutane

2-Bromobutane is an isomer of 1-bromobutane. Both compounds share the molecular formula C_4H_9Br. 2-Bromobutane is also known as sec-butyl bromide or methylethylbromomethane.

Asymmetric carbon

The term Asymmetric carbon is an expression used, mainly in older literature, for what is now more commonly called a stereogenic carbon. This expression refers to a carbon atom that is attached to four different atoms or four different groups of atoms. Knowing the number of asymmetric (stereogenic) carbon atoms, one can calculate the maximum possible number of enantiomers for any given molecule as follows:

If n is the number of stereogenic carbon atoms then the maximum number of isomers = 2^n

As an example, the molecule shown here has a stereogenic carbon represented by the C in the center of the structure: .

Bromobenzene

Bromobenzene s are a group of halobenzenes formed in a substitution reaction between bromine and benzene with a hydrogen bromide by-product. The name strictly refers to mono Bromobenzene , a benzene with a single bromine; however it can be used to refer to a benzene containing any number of bromine molecules. Bromobenzene is a clear pale yellow liquid.

Radicals

In chemistry, Radicals are atoms, molecules, or ions with unpaired electrons on an otherwise open shell configuration. These unpaired electrons are usually highly reactive, so Radicals are likely to take part in chemical reactions. Radicals play an important role in combustion, atmospheric chemistry, polymerization, plasma chemistry, biochemistry, and many other chemical processes, including human physiology.

Alcohol

In chemistry, an Alcohol is any organic compound in which a hydroxyl group (-OH) is bound to a carbon atom of an alkyl or substituted alkyl group. The general formula for a simple acyclic Alcohol is $C_nH_{2n+1}OH$. In common terms, the word Alcohol refers to ethanol, the type of Alcohol found in Alcohol ic beverages.

Ethanol is a colorless, volatile liquid with a mild odor which can be obtained by the fermentation of sugars.

Butyronitrile

Butyronitrile, or propyl cyanide, is a low molecular weight nitrile with the molecular formula C_4H_7N. It is a clear liquid that is miscible with ethanol, diethyl ether and dimethylformamide.

Butyronitrile can be prepared by the controlled cyanation of n-butanol with ammonia at 300 °C with Ni-Al_2O_3 catalysts.

Configuration

The configuration of a molecule is the permanent geometry that results from the spatial arrangement of its bonds. The ability of the same set of atoms to form two or more molecules with different configurations is stereoisomerism. configuration is distinct from chemical conformation, a shape attainable by bond rotations.

Carvone

Carvone is a member of a family of chemicals called terpenoids. Carvone is found naturally in many essential oils, but is most abundant in the oils from seeds of caraway (Carum carvi) and dill.

Carvone forms two mirror image forms or enantiomers: S-(+)-Carvone smells like caraway.

Fischer projection

The Fischer projection, devised by Hermann Emil Fischer in 1891, is a two-dimensional representation of a three-dimensional organic molecule by projection. They are used by chemists, particularly in organic chemistry and biochemistry. All bonds are depicted as horizontal or vertical lines.

Hormone

Hormone s are chemicals released by cells that affect cells in other parts of the body. Only a small amount of Hormone is required to alter cell metabolism. It is essentially a chemical messenger that transports a signal from one cell to another.

Sodium

Sodium is a metallic element with a symbol Na and atomic number 11. It is a soft, silvery-white, highly reactive metal and is a member of the alkali metals within 'group 1' (formerly known as 'group IA'.) It has only one stable isotope, ^{23}Na.

Hollywood

A Hollywood is a drug slang term referring to an unusually large and often dangerous single dosage of cocaine.

Depending on a user's experience with and tolerance of the drug, a typical line measures between 50-75mg of cocaine. However, a Hollywood line can exceed 200 mg of cocaine in a single dosage.

Enzymes

Enzymes are biomolecules that catalyze (i.e., increase the rates of) chemical reactions. Nearly all known Enzymes are proteins. However, certain RNA molecules can be effective biocatalysts too.

Epinephrine

Epinephrine is a hormone and neurotransmitter that participates in the 'fight or flight' response of the sympathetic nervous system. It is a catecholamine, a sympathomimetic monoamine produced by the adrenal glands from the amino acids phenylalanine and tyrosine.

The term Epinephrine is derived from the Greek roots epi- and nephros, and literally means on the kidney, in reference to the gland's anatomic location.

Racemic mixture

In chemistry, a Racemic mixture is one that has equal amounts of left- and right-handed enantiomers of a chiral molecule. The first known Racemic mixture was 'racemic acid,' which Louis Pasteur found to be a mixture of the two enantiomeric isomers of tartaric acid.

A racemate is optically inactive, meaning that it does not rotate plane-polarized light.

Enantiomeric excess

The Enantiomeric excess of a substance is a measure of how pure it is. In this case, the impurity is the undesired enantiomer (the 'opposite-handed' mirror image of a chiral compound.)

Enantiomeric excess is defined as the absolute difference between the mole fraction of each enantiomer:

$$ee = |F_+ - F_-|$$

where

$$F_+ + F_- = 1$$

In practice, it is most often expressed as a percent Enantiomeric excess.

Butane

Butane also called n Butane is the unbranched alkane with four carbon atoms, $CH_3CH_2CH_2CH_3$. Butane is also used as a collective term for n Butane together with its only other isomer, iso Butane (also called methylpropane), $CH(CH_3)_3$.

Butane s are highly flammable, colorless, odorless, easily liquefied gases.

Biphenyl

Biphenyl is an organic compound that forms colorless crystals. It has a distinctively pleasant smell. Biphenyl is an aromatic hydrocarbon with a molecular formula $(C_6H_5)_2$.

Allene

An Allene is a hydrocarbon in which one atom of carbon is connected by double bonds with two other atoms of carbon. Allene also is the common name for the parent compound of this series, propadiene.

Such pair of bonds make Allene s much more reactive than other alkenes.

Lactic acid

Lactic acid, a.k.a. Milk Acid, is a chemical compound that plays a role in several biochemical processes. It was first isolated in 1780 by a Swedish chemist, Carl Wilhelm Scheele, and is a carboxylic acid with a chemical formula of $C_3H_6O_3$. It has a hydroxyl group adjacent to the carboxyl group, making it an alpha hydroxy acid (AHA.)

Propane

Propane is a three-carbon alkane, normally a gas, but compressible to a transportable liquid. It is derived from other petroleum products during oil or natural gas processing. It is commonly used as a fuel for engines, oxy-gas torches, barbecues, portable stoves and residential central heating.

2-butene

2-Butene is a compound with formula C_4H_8. It is one of the isomers of butene. It is a good example of cis-trans isomerism.

Diastereomers

Diastereomers are stereoisomers that are not enantiomers (non-superimposable mirror images of each other.) Diastereomers can have different physical properties and different reactivity. In another definition Diastereomers are pairs of isomers that have opposite configurations at one or more of the chiral centers but are not mirror images of each other .

Dimethyl carbonate

Dimethyl carbonate, often abbreviated DMC, is a flammable clear liquid boiling at 90 °C. It has recently found use as a methylating reagent. Its main benefit over other methylating reagents such as iodomethane and dimethyl sulfate is its lesser toxicity and its biodegradability. Also, it is now prepared from catalytic oxidative carbonylation of methanol with carbon monoxide and oxygen, instead of from phosgene making its production non-toxic and environmentally friendly.

Alkene

In organic chemistry, an Alkene olefin, or olefine is an unsaturated chemical compound containing at least one carbon-to-carbon double bond. The simplest acyclic Alkene s, with only one double bond and no other functional groups, form a homologous series of hydrocarbons with the general formula C_nH_{2n}.

The simplest Alkene is ethylene (C_2H_4), which has the International Union of Pure and Applied Chemistry (IUPAC) name ethene.

Norepinephrine

Noradrenaline (BAN) or Norepinephrine is a catecholamine with dual roles as a hormone and a neurotransmitter.

	As a stress hormone, Norepinephrine affects parts of the brain where attention and responding actions are controlled. Along with epinephrine, Norepinephrine also underlies the fight-or-flight response, directly increasing heart rate, triggering the release of glucose from energy stores, and increasing blood flow to skeletal muscle.
Stereochemistry	Stereochemistry, a subdiscipline of chemistry, involves the study of the relative spatial arrangement of atoms within molecules. An important branch of Stereochemistry is the study of chiral molecules . Stereochemistry is a hugely important facet of chemistry and the study of stereochemical problems spans the entire range of organic, inorganic, biological, physical and supramolecular chemistries.
Meso compound	A Meso compound or meso isomer is a non-optically active member of a set of stereoisomers, at least two of which are optically active. This means that despite containing two or more stereocenters (chiral centers) it is not chiral. A Meso compound is superposable on its mirror image, and it does not produce a '(+)' or '(-)' reading when analyzed with a polarimeter.
Tartaric acid	Tartaric acid is a white crystalline diprotic organic acid. It occurs naturally in many plants, particularly grapes, bananas, and tamarinds, and is one of the main acids found in wine. It is added to other foods to give a sour taste, and is used as an antioxidant.
Glutamic acid	Glutamic acid is one of the 20 proteinogenic amino acids, and its codons are Glutamic acid A and Glutamic acid G. It is a non-essential amino acid. The carboxylate anions and salts of Glutamic acid are known as glutamates. The side chain carboxylic acid functional group has pK_a of 4.1 and exists in its negatively charged deprotonated carboxylate form at physiological pH. Although they occur naturally in many foods, the flavor contributions made by Glutamic acid and other amino acids were only scientifically identified early in the twentieth century.
Galactose	Galactose is a type of sugar which is less sweet than glucose. It is considered a nutritive sweetener because it has food energy. Its name comes from the Ancient Greek word for milk, γÎ¬λακτος .
Monosaccharides	Monosaccharides are the most basic unit of carbohydrates. They are the simplest form of sugar and are usually colorless, water-soluble, crystalline solids. Some Monosaccharides have a sweet taste.
Sugar	Sugar is a class of edible crystalline substances, mainly sucrose, lactose, and fructose. Human taste buds interpret its flavor as sweet. Sugar as a basic food carbohydrate primarily comes from Sugar cane and from Sugar beet, but also appears in fruit, honey, sorghum, Sugar maple (in maple syrup), and in many other sources.

Alkyl	In chemistry, an Alkyl group is a hydrocarbon; typically an Alkyl is a part of a larger molecule. The term is usually used loosely, there is no general formula for an Alkyl group. In structural formulae, an Alkyl group is represented with an R. Usually, Alkyl groups resemble hydrocarbons, but with one less hydrogen atom.
Aryl	In the context of organic molecules, Aryl refers to any functional group or substituent derived from a simple aromatic ring, may it be phenyl, thiophenyl, indolyl, etc . 'Aryl' is used for the sake of abbreviation or generalization. A simple Aryl group is phenyl, C_6H_5; it is derived from benzene.
Halogenoarene	In organic chemistry, a Halogenoarene, haloarene is an organic compound in which a halogen atom is bonded to a carbon atom which is part of an aromatic ring. The haloarene are studied separately from haloalkanes because they exhibit a lot of differences in methods of preparation and properties. There are two main preparatory routes for aryl halides.
Halide	A Halide is a binary compound, of which one part is a halogen atom and the other part is an element or radical that is less electronegative than the halogen, to make a fluoride, chloride, bromide, iodide, or astatide compound. Many salts are Halide s. All Group 1 metals form Halide s with the halogens and they are white solids.
Halothane	Halothane vapour (or Fluothane) is an inhalational general anaesthetic. Its IUPAC name is 2-bromo-2-chloro-1,1,1-trifluoroethane. It is the only inhalational anaesthetic agent containing a bromine atom; there are several other halogenated anesthesia agents which lack the bromine atom and do contain the fluorine and chlorine atoms present in Halothane.
Tetrafluoroethylene	Tetrafluoroethylene is a chemical compound with the formula C_2F_4. It is the simplest alkene fluorocarbon. This gaseous species is used primarily in the industrial preparation of polymers.
Thyroxine	Thyroxine, or 3,5,3',5'-tetraiodothyronine , a form of thyroid hormones is the major hormone secreted by the follicular cells of the thyroid gland. Thyroxine is synthesized via the iodination and covalent bonding of the phenyl portions of tyrosine residues found in an initial peptide, thyroglobulin, which is secreted into thyroid granules. These iodinated diphenyl compounds are cleaved from their peptide backbone upon being stimulated by thyroid stimulating hormone.
Vinyl Halide	In organic chemistry, a Vinyl halide is any alkene with at least one halide substituent bonded directly on one of the unsaturated carbons. Vinyl chloride is one such substance. Vinyl halides are very useful synthetic intermediates due to the vast number of reactions that make use of them.
Alcohol	In chemistry, an Alcohol is any organic compound in which a hydroxyl group (-OH) is bound to a carbon atom of an alkyl or substituted alkyl group. The general formula for a simple acyclic Alcohol is $C_nH_{2n+1}OH$. In common terms, the word Alcohol refers to ethanol, the type of Alcohol found in Alcohol ic beverages. Ethanol is a colorless, volatile liquid with a mild odor which can be obtained by the fermentation of sugars.

Halomethane

Halomethane compounds are molecules of methane (CH_4) with one or more of the hydrogen atoms replaced with halogen atoms. The halogens are found in group 17 of the periodic table of the elements. Forming covalent bonds, these compounds are generally stable.

Vicinal

In chemistry Vicinal stands for any two functional groups bonded to two adjacent carbon atoms. For example the molecule 2,3-dibromobutane carries two Vicinal bromine atoms and 1,3-dibromobutane does not.

Likewise in a gem-dibromide the prefix gem, an abbreviation of geminal, signals that both bromine atoms are bonded to the same atom.

Geminal

In chemistry, the term Geminal refers to the relationship between two functional groups that are attached to the same atom. The prefix gem is applied to a chemical name to denote this relationship, as in a gem-dibromide.

The following example shows the conversion of a cyclohexyl methyl ketone to a gem-dichloride through a reaction with phosphorus pentachloride.

Carbon

Carbon is the chemical element with symbol C and atomic number 6. As a member of group 14 on the periodic table, it is nonmetallic and tetravalent--making four electrons available to form covalent chemical bonds. There are three naturally occurring isotopes, with ^{12}C and ^{13}C being stable, while ^{14}C is radioactive, decaying with a half-life of about 5730 years.

Carbon tetrachloride

Carbon tetrachloride, also known by many other names is the organic compound with the formula CCl_4. It is a reagent in synthetic chemistry and was formerly widely used in fire extinguishers, as a precursor to refrigerants, and as a cleaning agent. It is a colourless liquid with a 'sweet' smell that can be detected at low levels.

Phosgene

Phosgene is the chemical compound with the formula $COCl_2$. This colorless gas gained infamy as a chemical weapon during World War I, but it is also a valued industrial reagent and building block in organic synthesis. In low concentrations, its odor resembles freshly cut hay or grass.

Reagent

A Reagent or reactant is a substance or compound consumed during a chemical reaction. Solvents and catalysts, although they are involved in the reaction, are usually not referred to as reactants.

Although the terms reactant and Reagent are often used interchangeably, a Reagent is more specifically 'a test substance that is added to a system in order to bring about a reaction or to see whether a reaction occurs'.

Kepone

Kepone is a carcinogenic insecticide related to mirex, used between 1966 and 1975 in the USA for ant and roach baits. It was produced by Allied Signal Company in Hopewell, Virginia and produced nationwide pollution controversy due to improper handling and dumping of the substance into the James River. Its use was banned in 1975.

Ether

Ether is a class of organic compounds which contain an Ether group -- an oxygen atom connected to two (substituted) alkyl or aryl groups -- of general formula R-O-R'. A typical example is the solvent and anesthetic diethyl Ether, commonly referred to simply as 'Ether' (ethoxyethane, CH_3-CH_2-O-CH_2-CH_3.)

Ether molecules cannot form hydrogen bonds amongst each other, resulting in a relatively low boiling point compared to that of the analogous alcohols.

Iodide

An Iodide ion is an iodine atom with a −1 charge. Compounds with iodine in formal oxidation state −1 are called Iodide s. This can include ionic compounds such as caesium Iodide or covalent compounds such as phosphorus tri Iodide .

Cyclohexane

Cyclohexane is a cycloalkane with the molecular formula C_6H_{12}. Cyclohexane is used as a nonpolar solvent for the chemical industry, and also as a raw material for the industrial production of adipic acid and caprolactam, both of which are intermediates used in the production of nylon. On an industrial scale, Cyclohexane is produced by reacting benzene with hydrogen.

Halogenation

Halogenation is a chemical reaction that incorporates a halogen atom into a molecule. More specific descriptions exist that specify the type of halogen: fluorination, chlorination, bromination, and iodination.

In a Markovnikov addition reaction, a halogen like bromine is reacted with an alkene which causes the π-bond to break forming an haloalkane.

Isobutane

Isobutane is an alkane, isomeric with butane. Recent concerns with depletion of the ozone layer by freon gases have led to increased use of Isobutane as a gas for refrigeration systems, especially in domestic refrigerators and freezers, and as a propellant in aerosol sprays. When used as a refrigerant or a propellant, Isobutane is also known as R-600a.

Alkanes

Alkanes are chemical compounds that consist only of the elements carbon and hydrogen, wherein these atoms are linked together exclusively by single bonds without any cyclic structure Alkanes belong to a homologous series of organic compounds in which the members differ by a constant relative atomic mass of 14.

Each carbon atom must have 4 bonds, and each hydrogen atom must be joined to a carbon atom

Alkynes

Alkynes are hydrocarbons that have a triple bond between two carbon atoms, with the formula C_nH_{2n-2}. Alkynes are traditionally known as acetylenes, although the name acetylene also refers specifically to C_2H_2, known formally (but rarely) as ethyne using IUPAC nomenclature. Like other hydrocarbons, Alkynes are generally hydrophobic but tend to be more reactive.

Propane

Propane is a three-carbon alkane, normally a gas, but compressible to a transportable liquid. It is derived from other petroleum products during oil or natural gas processing. It is commonly used as a fuel for engines, oxy-gas torches, barbecues, portable stoves and residential central heating.

Cyclohexene

Cyclohexene is a colorless clear liquid cycloalkene with an intense aversive characteristic sharp smell reminiscent of an oil refinery.

It is not very stable upon long term storage with exposure to light and air and should be distilled before use to eliminate peroxides. A common experiment among beginning organic chemistry students is the acid catalyzed dehydration of cyclohexanol with distillative removal of the resulting Cyclohexene from the reaction mixture:

- Refractive Index: 1.4465
- Critical temperature: 287.2 °C (560.4 K)

Dehydrohalogenation

Dehydrohalogenation is an organic reaction from which an alkene is obtained from an alkyl halide . It is also called a β-Elimination reaction and is a type of elimination reaction. Ethanolic potassium hydroxide when reacted with alkyl halide gives alkene.

Leaving group

A Leaving group in chemistry is an ion or substituent with the ability to detach itself from a molecule. The remaining molecule or fragment remaining is known as the residual or main part. The term Leaving group is dependent on the context of the statement.

Nucleophilic substitution

In organic and inorganic chemistry, Nucleophilic substitution is a fundamental class of substitution reaction in which an 'electron rich' nucleophile selectively bonds with or attacks the positive or partially positive charge of an atom attached to a group or atom called the leaving group; the positive or partially positive atom is referred to as an electrophile.

The most general form for the reaction may be given as

Nuc: + R-LG → R-Nuc + LG:

The electron pair (:) from the nucleophile (Nuc) attacks the substrate (R-LG) forming a new bond, while the leaving group (LG) departs with an electron pair. The principal product in this case is R-Nuc.

Nucleophile

In chemistry, a Nucleophile is a reagent that forms a chemical bond to its reaction partner (the electrophile) by donating both bonding electrons. Because nucleophiles donate electrons, they are by definition Lewis bases All molecules or ions with a free pair of electrons can act as nucleophiles.

Substrate

In biochemistry, a substrate is a molecule upon which an enzyme acts. Enzymes catalyze chemical reactions involving the substrate(s.) In the case of a single substrate, the substrate binds with the enzyme active site, and an enzyme-substrate complex is formed.

Concerted reaction

In chemistry, a Concerted reaction is a chemical reaction in which all bond breaking and bond making occurs in a single step. Reactive intermediates or other unstable high energy intermediates are not involved. Concerted reaction rates tend not to depend on solvent polarity ruling out large buildup of charge in the transition state.

Methanol

Methanol, also known as methyl alcohol, carbinol, wood alcohol, wood naphtha or wood spirits, is a toxic chemical with chemical formula CH_3OH Drinking even small amounts can cause blindness. It is the simplest alcohol, and is a light, volatile, colourless, flammable, toxic liquid with a distinctive odor that is very similar to but slightly sweeter than ethanol (drinking alcohol.)

Methoxide

Methoxide is an organic salt, and the simplest alkoxide. In organic chemistry, the Methoxide ion has a formula of CH_3O^- and is the conjugate base of methanol.

Sodium Methoxide, also referred to as Sodium methylate, is a white powder when pure.

Nitrile

A Nitrile is any organic compound which has a -C≡N functional group. The -C≡N functional group is called a Nitrile group. In the -CN group, the carbon atom and the nitrogen atom are triple bonded together.

Strain

In chemistry a molecule experiences Strain when in a chemical conformation there exist unfavorable bond angles or bond distances. Strain energy is released when the molecule can relax to a conformation with less Strain or when the molecule interacts in a suitable chemical reaction.

Several types of Strain exist:

- Angle Strain
- Torsional Strain
- van der Waals Strain

.

Acetone

Acetone is the organic compound with the formula $OC(CH_3)_2$. This colorless, mobile, flammable liquid is the simplest example of the ketones. Owing to the fact that Acetone is miscible with water, and it serves as an important solvent in its own right, typically the solvent of choice for cleaning purposes in the laboratory.

Acetonitrile

Acetonitrile is the chemical compound with formula CH_3CN. This colourless liquid is the simplest organic nitrile. It is produced mainly as a byproduct of acrylonitrile manufacture. It is widely used as a polar aprotic solvent in synthetic chemistry, and as a medium-polarity solvent in HPLC.

Acetonitrile is a by-product from the manufacture of acrylonitrile.

Crown ethers

Crown ethers are heterocyclic chemical compounds that consist of a ring containing several ether groups. The most common Crown ethers are oligomers of ethylene oxide, the repeating unit being ethyleneoxy, i.e., $-CH_2CH_2O-$. Important members of this series are the tetramer (n = 4), the pentamer (n = 5), and the hexamer (n = 6.)

Isopropyl

In organic chemistry, Isopropyl is a substituent form of propane, the three-carbon alkyl functional group. As an isomer of propyl, bonding to an R group occurs at the secondary (2°) carbon.

Isopropyl is also known as 1-methylethyl (IUPAC); shorthand notations for this alkyl are i-Pr,^{i}Pr, or Pr^{i} (some care has to be taken when using the abbreviation 'Pr', as it is the same as for the chemical element Praseodymium.)

Bromide

A Bromide ion is a bromine atom with charge of −1.

Compounds with bromine in formal oxidation state −1 are called Bromide s, and each individual chemical in this class can be called a Bromide as well. The class name can include ionic compounds such as caesium Bromide or covalent compounds such as sulfur di Bromide .

Configuration

The configuration of a molecule is the permanent geometry that results from the spatial arrangement of its bonds. The ability of the same set of atoms to form two or more molecules with different configurations is stereoisomerism. configuration is distinct from chemical conformation, a shape attainable by bond rotations.

Fischer projection

The Fischer projection, devised by Hermann Emil Fischer in 1891, is a two-dimensional representation of a three-dimensional organic molecule by projection. They are used by chemists, particularly in organic chemistry and biochemistry. All bonds are depicted as horizontal or vertical lines.

Walden inversion	Walden inversion is the inversion of a chiral center in a molecule in a chemical reaction. Since a molecule can form two enantiomers around a chiral center, the Walden inversion converts the configuration of the molecule from one enantiomeric form to the other. For example, in a S_N2 reaction, Walden inversion occurs at a tetrahedral carbon atom.
Stereochemistry	Stereochemistry, a subdiscipline of chemistry, involves the study of the relative spatial arrangement of atoms within molecules. An important branch of Stereochemistry is the study of chiral molecules . Stereochemistry is a hugely important facet of chemistry and the study of stereochemical problems spans the entire range of organic, inorganic, biological, physical and supramolecular chemistries.
Isomer	In chemistry, Isomer s are compounds with the same molecular formula but different structural formula. Isomer s do not necessarily share similar properties unless they also have the same functional groups. This should not be confused with a nuclear Isomer which involves a nucleus at different states of excitement.
Solvolysis	Solvolysis is a special type of nucleophilic substitution or elimination where the nucleophile is a solvent molecule. For certain nucleophiles, there are specific terms for the type of Solvolysis reaction. For water, the term is hydrolysis; for alcohols, it is alcoholysis; for ammonia, it is ammonolysis.
Allyl	An Allyl group is an alkene hydrocarbon group with the formula $H_2C{=}CH{-}CH_2{-}$. It is made up of a vinyl group, $CH_2{=}CH{-}$, attached to a methylene $-CH_2$. For example Allyl alcohol has the structure $H_2C{=}CH{-}CH_2OH$. Another example of a simple Allyl compound is Allyl chloride.
Allyl bromide	Allyl bromide is an organic halide. Its refractive index is 1.4697 (20 °C, 589 nm.) Allyl bromide is an alkylating agent used in synthesis of polymers, pharmaceuticals, allyls and other organic compounds.
Substituent	In organic chemistry and biochemistry, a Substituent is an atom or group of atoms substituted in place of a hydrogen atom on the parent chain of a hydrocarbon. The terms Substituent, side chain, group, branch, or pendant group are used almost interchangeably to describe branches from a parent structure, though certain distinctions are made in the context of polymer chemistry. In polymers, side chains extend from a backbone structure.
Alkene	In organic chemistry, an Alkene olefin, or olefine is an unsaturated chemical compound containing at least one carbon-to-carbon double bond. The simplest acyclic Alkene s, with only one double bond and no other functional groups, form a homologous series of hydrocarbons with the general formula C_nH_{2n}. The simplest Alkene is ethylene (C_2H_4), which has the International Union of Pure and Applied Chemistry (IUPAC) name ethene.
Racemization	In chemistry Racemization refers to partial conversion of one enantiomer into another. Chiral molecules have two forms (at each point of asymmetry) which differ in their optical characteristics: the levorotatory form (the (−)-form) will rotate the plane of polarization of a beam of light to the left, while the dextrorotatory form (the (+)-form) will rotate the plane of polarization of a beam of light to the right. The two forms, which are non-superimposable when rotated in 3 dimensional space, are said to be diastereomers.
Cicutoxin	Cicutoxin is a poisonous polyyne and alcohol found in various plants, most notably water hemlock (Cicuta species.) It is structurally related to the oenanthotoxin of hemlock water dropwort.

	It is a cholinergic poison, causing death by disruption of the central nervous system.
Camphor	Camphor is a waxy, white or transparent solid with a strong, aromatic odor. It is a terpenoid with the chemical formula $C_{10}H_{16}O$. It is found in wood of the Camphor laurel (Cinnamomum Camphor a), a large evergreen tree found in Asia It also occurs in some other related trees in the laurel family, notably Ocotea usambarensis.
Terpene	Terpenes are a large and varied class of hydrocarbons, produced primarily by a wide variety of plants, particularly conifers, though also by some insects such as termites or swallowtail butterflies, which emit terpenes from their osmeterium. They are the major components of resin, and of turpentine produced from resin. The name 'Terpene' is derived from the word 'turpentine'.
Nitrate	In inorganic chemistry, a Nitrate is a salt of nitric acid with an ion composed of one nitrogen and three oxygen atoms (NO_3^-.) In organic chemistry the esters of nitric acid and various alcohols are called nitrates. The Nitrate ion is a polyatomic ion with the empirical formula NO_3^- and a molecular mass of 62.0049.
Dehydration reaction	In chemistry, a Dehydration reaction is usually defined as a chemical reaction that involves the loss of water from the reacting molecule. Dehydration reaction s are a subset of elimination reactions. Because the hydroxyl group (-OH) is a poor leaving group, having an Brønsted acid catalyst often helps by protonating the hydroxyl group to give the better leaving group, $-OH_2^+$.
Carbocation	A Carbocation is an ion with a positively-charged carbon atom. The charged carbon atom in a Carbocation is a 'sextet', i.e. it has only six electrons in its outer valence shell instead of the eight valence electrons that ensures maximum stability . Therefore Carbocation s are often reactive, seeking to fill the octet of valence electrons as well as regain a neutral charge.
1-bromobutane	1-Bromobutane$_2CH_2Br$) is a colorless liquid that is insoluble in water, but soluble in ethanol and diethyl ether. As a primary alkyl halide, it is especially prone to S_N2 type reactions. It is commonly used as an alkylating agent, or in combination with magnesium metal in dry ether (Grignard reagent) to form carbon-carbon bonds.
2-bromobutane	2-Bromobutane is an isomer of 1-bromobutane. Both compounds share the molecular formula C_4H_9Br. 2-Bromobutane is also known as sec-butyl bromide or methylethylbromomethane.
Bromobenzene	Bromobenzene s are a group of halobenzenes formed in a substitution reaction between bromine and benzene with a hydrogen bromide by-product. The name strictly refers to mono Bromobenzene , a benzene with a single bromine; however it can be used to refer to a benzene containing any number of bromine molecules. Bromobenzene is a clear pale yellow liquid.
Sodium	Sodium is a metallic element with a symbol Na and atomic number 11. It is a soft, silvery-white, highly reactive metal and is a member of the alkali metals within 'group 1' (formerly known as 'group IA'.) It has only one stable isotope, ^{23}Na.
Sodium ethoxide	Sodium ethoxide is an alkoxide salt with the chemical formula C_2H_5ONa.

It is commercially available as a dry yellow solid, or as a solution in ethanol. It is easily prepared in the laboratory by reacting sodium metal with ethanol:

$$2\ C_2H_5OH + 2\ Na \rightarrow 2\ C_2H_5ONa + H_2$$

Sodium ethoxide is commonly used in the Claisen condensation and malonic ester synthesis, if an ethyl ester is one of the reactants.

Chirality

In general, chiral molecules have point Chirality at a single stereogenic atom, usually carbon, which has four different substituents. The two enantiomers of such compounds are said to have different absolute configurations at this center. This center is thus stereogenic (i.e., a grouping within a molecular entity that may be considered a focus of stereoisomerism), and is exemplified by the α-carbon of amino acids.

McLafferty rearrangement

The McLafferty rearrangement is a reaction observed in mass spectrometry. It is sometimes found that a molecule containing a keto-group undergoes β-cleavage, with the gain of the γ-hydrogen atom. This rearrangement may take place by a radical or ionic mechanism.

Aldehyde

An Aldehyde is an organic compound containing a terminal carbonyl group. This functional group, which consists of a carbon atom bonded to a hydrogen atom and double-bonded to an oxygen atom (chemical formula O=CH-), is called the Aldehyde group. The Aldehyde group is also called the formyl or methanoyl group.

Enantiomer

In chemistry, an Enantiomer is one of two stereoisomers that are non-superposable complete mirror images of each other, much as one's left and right hands are 'the same' but opposite. Enantiopure compounds refer to a sample having within the limits of detection, molecules of only one chirality.

Enantiomer s have, when present in a symmetric environment, identical chemical and physical properties except for their ability to rotate plane-polarized light (+/-) by equal amounts but in opposite directions.

Absolute configuration

An Absolute configuration in stereochemistry is the spatial arrangement of the atoms of a chiral molecular entity (or group) and its stereochemical description e.g. R or S.

Absolute configuration s for chiral molecules are traditionally obtained by X-ray crystallography but only when the compound crystallises in one of the 65 Sohncke Groups (Chiral Space Groups.) Alternative techniques are Optical rotatory dispersion, vibrational circular dichroism and the use of chiral shift reagents in proton NMR.

In step two the assignment of R or S is based on the Cahn-Ingold-Prelog priority rules.

Absolute configuration s are also relevant to characterization of crystals.

Williamson ether synthesis

The Williamson ether synthesis is an organic reaction, forming an ether from a organohalide and an alcohol. This reaction was developed by Alexander Williamson in 1850 . Typically it involves the reaction of an alkoxide ion with a primary alkyl halide via an S_N2 reaction.

Amine	Amine s are organic compounds and functional groups that contain a basic nitrogen atom with a lone pair. Amine s are derivatives of ammonia, wherein one or more hydrogen atoms have been replaced by a substituent such as an alkyl or aryl group. Important Amine s include amino acids, biogenic Amine s, trimethyl Amine , and aniline; see Category: Amine s for a list of Amine s.
Organolithium reagent	An Organolithium reagent is an organometallic compound with a direct bond between a carbon and a lithium atom. As the electropositive nature of lithium puts most of the charge density of the bond on the carbon atom, effectively creating a carbanion, organolithium compounds are extremely powerful bases and nucleophiles. Organolithium reagents are industrially prepared by the reaction of an halocarbon with lithium metal, i.e. R-X + 2 Li → R-Li + LiX. A side reaction of this synthesis, especially with alkyl iodides, is the Wurtz reaction, in which an R-Li species reacts with an R-X species forming an R-R coupled product.

Alkene

In organic chemistry, an Alkene olefin, or olefine is an unsaturated chemical compound containing at least one carbon-to-carbon double bond. The simplest acyclic Alkene s, with only one double bond and no other functional groups, form a homologous series of hydrocarbons with the general formula C_nH_{2n}.

The simplest Alkene is ethylene (C_2H_4), which has the International Union of Pure and Applied Chemistry (IUPAC) name ethene.

Carbon-carbon bond

A Carbon-carbon bond is a covalent bond between two carbon atoms. The most common form is the single bond - a bond composed of two electrons, one from each of the two atoms. The carbon-carbon single bond is a sigma bond and is said to be formed between one hybridized orbital from each of the carbon atoms.

Ethylene

Ethylene is the chemical compound with the formula C_2H_4. It is the simplest alkene. Because it contains a carbon-carbon double bond, Ethylene is called an unsaturated hydrocarbon or an olefin.

Pinene

The chemical compound Pinene is a bicyclic terpene ($C_{10}H_{16}$, 136.24 g/mol) known as a monoterpene . There are two structural isomers found in nature: α-Pinene and β-Pinene. As the name suggests, both forms are important constituents of pine resin; they are also found in the resins of many other conifers, and more widely in other plants.

Turpentine

Turpentine is a fluid obtained by the distillation of resin obtained from trees, mainly pine trees. It is composed of terpenes, mainly the monoterpenes alpha-pinene and beta-pinene. It is sometimes known colloquially as turps, but this more often refers to Turpentine substitute

2-butene

2-Butene is a compound with formula C_4H_8. It is one of the isomers of butene. It is a good example of cis-trans isomerism.

Cyclopropane

Cyclopropane is a cycloalkane molecule with the molecular formula C_3H_6, consisting of three carbon atoms linked to each other to form a ring, with each carbon atom bearing two hydrogen atoms.

The bonds between the carbon atoms are considerably weaker than in a typical carbon-carbon bond, yielding reactivity similar to or greater than alkenes. Baeyer strain theory explains why: the angle strain from the 60° angle between the carbon atoms (less than the normal angle of 109.5° for bonds between atoms with sp^3 hybridised orbitals) reduces the compound's carbon-carbon bond energy, making it more reactive than other cycloalkanes such as cyclohexane and cyclopentane.

Propane

Propane is a three-carbon alkane, normally a gas, but compressible to a transportable liquid. It is derived from other petroleum products during oil or natural gas processing. It is commonly used as a fuel for engines, oxy-gas torches, barbecues, portable stoves and residential central heating.

Propene

Propene, also known as propylene, is an unsaturated organic compound having the chemical formula C_3H_6. It has one double bond, and is the second simplest member of the alkene class of hydrocarbons, and it is also second in natural abundance.

At room temperature, Propene is a colourless, odourless gas, though when used as a fuel, it is mixed with minute quantities of foul-smelling sulfurous compounds (mercaptans) so that gas leaks can be readily detected.

Isomer

In chemistry, Isomer s are compounds with the same molecular formula but different structural formula. Isomer s do not necessarily share similar properties unless they also have the same functional groups. This should not be confused with a nuclear Isomer which involves a nucleus at different states of excitement.

Saturation

In chemistry, saturation has five different meanings:

1. In physical chemistry, saturation is the point at which a solution of a substance can dissolve no more of that substance and additional amounts of it will appear as a precipitate. This point of maximum concentration, the saturation point, depends on the temperature of the liquid as well as the chemical nature of the substances involved. This can be used in the process of recrystallisation to purify a chemical: it is dissolved to the point of saturation in hot solvent, then as the solvent cools and the solubility decreases, excess solute precipitates. Impurities, being present in much lower concentration, do not saturate the solvent and so remain dissolved in the liquid. If a change in conditions (e.g. cooling) means that the concentration is actually higher than the saturation point, the solution has become supersaturated.
2. In physical chemistry, when referring to surface processes, saturation denotes the degree of which a binding site is fully occupied. For example, base saturation refers to the fraction of exchangeable cations that are base cations. Similarly, in environmental soil science, nitrogen saturation means that an ecosystem, such as a soil, cannot store anymore nitrogen.
3. In organic chemistry, a saturated compound has no double or triple bonds. In saturated linear hydrocarbons, every carbon atom is attached to two hydrogen atoms, except those at the ends of the chain, which bear three hydrogen atoms. In the case of saturated methane, four hydrogen atoms are attached to the single, central carbon atom. Of simple hydrocarbons, alkanes are saturated, and alkenes are unsaturated. The degree of unsaturation specifies the amount of hydrogen that a compound can bind. The term is applied similarly to the fatty acid constituents of lipids, where the fat is described as saturated or unsaturated, depending on whether the constituent fatty acids contain carbon-carbon double bonds. Unsaturated is used when any carbon structure contains double or occasionally triple bonds. Many vegetable oils contain fatty acids with one (monounsaturated) or more (polyunsaturated) double bonds in them. The bromine number is an index of unsaturation.
4. In organometallic chemistry, an unsaturated complex has fewer than 18 valence electrons and thus is susceptible to oxidative addition or coordination of an additional ligand. Unsaturation is characteristic of many catalysts because it is usually a requirement for substrate activation.
5. In biochemistry, the term saturation refers to the fraction of total protein binding sites that are occupied at any given time.

.

1-butene

1-Butene is an organic compound and one of the isomers of butene. The formula is C_4H_8.

1-Butene is stable in itself but polymerizes exothermically.

Cyclobutane

Cyclobutane C_4H_8, with a molecular mass of 56.107g/mol, is a four carbon alkane in which all the carbon atoms are arranged cyclically, hence Cyclobutane Cyclobutane is a gas and commercially available as a liquefied gas.

Cyclobutane s are Cyclobutane derivatives.

Heteroatom

In the nomenclature of organic chemistry, a Heteroatom is any atom that is not carbon or hydrogen. It is typically, but not exclusively, nitrogen, oxygen, sulfur, phosphorus, boron, chlorine, bromine, or iodine.

In the description of protein structure, particularly in the now-deprecated Protein Data Bank file format, a Heteroatom record (HETATM) describes an atom belonging to a small molecule cofactor rather than to part of a biopolymer chain.

Isobutylene

Isobutylene is a hydrocarbon of significant industrial importance. It is a four-carbon branched alkene (olefin), one of the four isomers of butylene. At standard temperature and pressure it is a colorless flammable gas.

Alkanes

Alkanes are chemical compounds that consist only of the elements carbon and hydrogen, wherein these atoms are linked together exclusively by single bonds without any cyclic structure Alkanes belong to a homologous series of organic compounds in which the members differ by a constant relative atomic mass of 14.

Each carbon atom must have 4 bonds, and each hydrogen atom must be joined to a carbon atom

Configuration

The configuration of a molecule is the permanent geometry that results from the spatial arrangement of its bonds. The ability of the same set of atoms to form two or more molecules with different configurations is stereoisomerism. configuration is distinct from chemical conformation, a shape attainable by bond rotations.

Alcohol

In chemistry, an Alcohol is any organic compound in which a hydroxyl group (-OH) is bound to a carbon atom of an alkyl or substituted alkyl group. The general formula for a simple acyclic Alcohol is $C_nH_{2n+1}OH$. In common terms, the word Alcohol refers to ethanol, the type of Alcohol found in Alcohol ic beverages.

Ethanol is a colorless, volatile liquid with a mild odor which can be obtained by the fermentation of sugars.

1-bromobutane

1-Bromobutane$_2CH_2Br$) is a colorless liquid that is insoluble in water, but soluble in ethanol and diethyl ether. As a primary alkyl halide, it is especially prone to S_N2 type reactions. It is commonly used as an alkylating agent, or in combination with magnesium metal in dry ether (Grignard reagent) to form carbon-carbon bonds.

2-bromobutane

2-Bromobutane is an isomer of 1-bromobutane. Both compounds share the molecular formula C_4H_9Br. 2-Bromobutane is also known as sec-butyl bromide or methylethylbromomethane.

3-pentanone

3-Pentanone is a colorless liquid ketone with an odor like that of acetone. Its formula is $C_5H_{10}O$. It is soluble in about 25 parts water, and miscible with ethanol and diethyl ether.

Two other ketones, which are isomers of 3-Pentanone, are 2-pentanone and methyl isopropyl ketone.

Bromobenzene

Bromobenzene s are a group of halobenzenes formed in a substitution reaction between bromine and benzene with a hydrogen bromide by-product. The name strictly refers to mono Bromobenzene , a benzene with a single bromine; however it can be used to refer to a benzene containing any number of bromine molecules. Bromobenzene is a clear pale yellow liquid.

Cyclohexene

Cyclohexene is a colorless clear liquid cycloalkene with an intense aversive characteristic sharp smell reminiscent of an oil refinery.

It is not very stable upon long term storage with exposure to light and air and should be distilled before use to eliminate peroxides. A common experiment among beginning organic chemistry students is the acid catalyzed dehydration of cyclohexanol with distillative removal of the resulting Cyclohexene from the reaction mixture:

- Refractive Index: 1.4465
- Critical temperature: 287.2 °C (560.4 K)

Dimethylformamide

Dimethylformamide is the organic compound with the formula $(CH_3)_2NC(O)H$. Commonly abbreviated DMF (though this acronym is sometimes used for dimethylfuran), this colourless liquid is miscible with water and the majority of organic liquids. DMF is a common solvent for chemical reactions. Pure Dimethylformamide is odorless whereas technical grade or degraded Dimethylformamide often has a fishy smell due to impurity of dimethylamine.

Toluene

Toluene phenylmethane, and Toluol, is a clear water-insoluble liquid with the typical smell of paint thinners, redolent of the sweet smell of the related compound benzene. It is an aromatic hydrocarbon that is widely used as an industrial feedstock and as a solvent. Like other solvents, Toluene is also used as an inhalant drug for its intoxicating properties; however this causes severe neurological harm.

Propylene oxide

Propylene oxide is an organic compound with the molecular formula CH_3CHCH_2O. This colourless volatile liquid is produced on a large scale industrially, its major application being its use for the production of polyether polyols for use in making polyurethane plastics. It is chiral epoxide, although it commonly used as a racemic mixture.

Industrial production of Propylene oxide starts from propylene.

Chirality

In general, chiral molecules have point Chirality at a single stereogenic atom, usually carbon, which has four different substituents. The two enantiomers of such compounds are said to have different absolute configurations at this center. This center is thus stereogenic (i.e., a grouping within a molecular entity that may be considered a focus of stereoisomerism), and is exemplified by the α-carbon of amino acids.

Enantiomer

In chemistry, an Enantiomer is one of two stereoisomers that are non-superposable complete mirror images of each other, much as one's left and right hands are 'the same' but opposite. Enantiopure compounds refer to a sample having within the limits of detection, molecules of only one chirality.

Enantiomer s have, when present in a symmetric environment, identical chemical and physical properties except for their ability to rotate plane-polarized light (+/-) by equal amounts but in opposite directions.

Allyl

An Allyl group is an alkene hydrocarbon group with the formula $H_2C=CH-CH_2$-. It is made up of a vinyl group, $CH_2=CH$-, attached to a methylene $-CH_2$. For example Allyl alcohol has the structure $H_2C=CH-CH_2OH$. Another example of a simple Allyl compound is Allyl chloride.

Allyl chloride

Allyl chloride is the organic compound with the formula $CH_2=CHCH_2Cl$. This colorless liquid is insoluble in water but soluble in common organic solvents. It is mainly converted to epichlorohydrin, used in the production of plastics.

Diastereomers

Diastereomers are stereoisomers that are not enantiomers (non-superimposable mirror images of each other.) Diastereomers can have different physical properties and different reactivity. In another definition Diastereomers are pairs of isomers that have opposite configurations at one or more of the chiral centers but are not mirror images of each other .

Methylene

Methylene is a chemical species in which a carbon atom is bonded to two hydrogen atoms. Three different possibilities present themselves:

- the $-CH_2-$ group, e.g. dichloromethane (also known as Methylene chloride)
- the $=CH_2$ group, e.g. methylenecyclopropene,
- and the $:CH_2$ molecule, a carbene known as Methylene. A carbene is a highly reactive organic molecule having a divalent carbon atom with six valence electrons.

Methylene groups in a chain or ring contribute to its size and lipophilicity.

Phenyl group

In organic chemistry, the Phenyl group or phenyl ring is the functional group with the formula

C_6H_5-,

where the six carbon atoms are arranged in a cyclic ring structure. This hydrophobic, highly-stable and aromatic hydrocarbon unit can be found in many organic compounds. It can be thought of as being derived from benzene .

Substituent

In organic chemistry and biochemistry, a Substituent is an atom or group of atoms substituted in place of a hydrogen atom on the parent chain of a hydrocarbon. The terms Substituent, side chain, group, branch, or pendant group are used almost interchangeably to describe branches from a parent structure, though certain distinctions are made in the context of polymer chemistry. In polymers, side chains extend from a backbone structure.

Vinyl halide

In organic chemistry, a Vinyl halide is any alkene with at least one halide substituent bonded directly on one of the unsaturated carbons. Vinyl chloride is one such substance.

Vinyl halides are very useful synthetic intermediates due to the vast number of reactions that make use of them.

Cis-trans isomerism

In organic chemistry, Cis-trans isomerism or geometric isomerism or configuration isomerism or E-Z isomerism is a form of stereoisomerism describing the orientation of functional groups within a molecule. In general, such isomers contain double bonds, which cannot rotate, but they can also arise from ring structures, wherein the rotation of bonds is greatly restricted.

The term 'geometric isomerism' is considered an obsolete synonym of 'Cis-trans isomerism' by IUPAC. It is sometimes used as a synonym for general stereoisomerism (e.g., optical isomerism being called geometric isomerism); the correct term for non-optical stereoisomerism is diastereomerism.

Polyolefin

A Polyolefin is a polymer produced from a simple olefin (also called an alkene with the general formula C_nH_{2n}) as a monomer. For example, polyethylene is the Polyolefin produced by polymerizing the olefin ethylene. An equivalent term is polyalkene; this is a more modern term, although Polyolefin is still used in the petrochemical industry.

Cracking	In petroleum geology and chemistry, Cracking is the process whereby complex organic molecules such as kerogens or heavy hydrocarbons are broken down into simpler molecules (e.g. light hydrocarbons) by the breaking of carbon-carbon bonds in the precursors. The rate of Cracking and the end products are strongly dependent on the temperature and presence of any catalysts. Cracking, also referred to as pyrolysis, is the breakdown of a large alkane into smaller, more useful alkanes and an alkene.
Hydrogenation	Hydrogenation is the chemical reaction that results from the addition of hydrogen (H_2.) The process is usually employed to reduce or saturate organic compounds. The process typically constitutes the addition of pairs of hydrogen atoms to a molecule.
2,2-dimethylbutane	2,2-Dimethylbutane, trivially known as neohexane, is the isomer of hexane containing a quaternary carbon.It is a hydrocarbon i.e it contains hydrogen and carbon only. It has covalent bonds and is single bonded.
Dimethylamine	Dimethylamine is an organic compound with the formula $(CH_3)_2NH$. This secondary amine is a colorless, flammable liquified gas with an ammonia- or fish-like odor. Dimethylamine is generally encountered as a solution in water at concentrations up to around 40%. In 2005, an estimated 270,000 tons were produced.
Cyclobutene	Cyclobutene is a cycloalkene with chemical formula C_4H_6 and CAS number 822-35-5. It is used in chemical industry as a monomer for synthesis of some polymers and for a range of chemical syntheses.
Cyclopentane	Cyclopentane is a highly flammable alicyclic hydrocarbon with chemical formula C_5H_{10} and CAS number 287-92-3, consisting of a ring of five carbon atoms each bonded with two hydrogen atoms above and below the plane. It occurs as a colorless liquid with a petrol-like odor. Its melting point is −94 °C and its boiling point is 49 °C. The typical structure of Cyclopentane is the 'envelope' conformation. Cyclopentane is used in the manufacture of synthetic resins and rubber adhesives and also as a blowing agent in the manufacture of polyurethane insulating foam, as found in may domestic appliances such as refrigerators and freezers, replacing environmentally damaging alternatives such as CFC-11 and HCFC-141b More advanced technologies, such as computer hard drives and outerspace equipment employ multiply-alkylated Cyclopentane lubricants because of their extremely low volatility.
Cyclopentene	Cyclopentene is a chemical compound with the formula C_5H_8. It is a colorless liquid with a petrol-like odor. It is one of the cycloalkenes.
Ring strain	Ring strain is an organic chemistry term that describes the destabilization of a cyclic molecule--such as a cycloalkane--due to the non-favorable high energy spatial orientations of its atoms. Non-cyclic molecules do not exhibit Ring strain because their terminal (end) atoms are not connected to force a particular type of spatial orientation. Ring strain results from a combination of angle strain, conformational strain or Pitzer strain, and transannular strain or van der Waals strain.
Cyclopropene	Cyclopropene is an organic compound with the formula C_3H_4. It is the simplest isolable cycloalkene. It has a triangular structure.

Transamination	There are two chemical reactions known as Transamination The first is the reaction between an amino acid and an alpha-keto acid. The amino group is transferred from the former to the latter; this results in the amino acid being converted to the corresponding α-keto acid, while the reactant α-keto acid is converted to the corresponding amino acid (if the amino group is removed from an amino acid, an α-keto acid is left behind.)
Cycloalkene	A Cycloalkene or cycloolefin is a type of alkene hydrocarbon which contains a closed ring of carbon atoms, but has no aromatic character. Some Cycloalkene s, such as cyclobutene and cyclopentene, can be used as monomers to produce polymer chains.
Decahydronaphthalene	Decahydronaphthalene, a bicyclic organic compound, is an industrial solvent. A colorless liquid with an aromatic odor, it is used as a solvent for many resins. It is the saturated analog of naphthalene and can be prepared from it by hydrogenation in a fused state in the presence of a catalyst.
Norbornane	Norbornane is an organic compound and a saturated hydrocarbon with chemical formula C_7H_{12}. It is a crystalline compound with melting point 88 °C. The carbon skeleton is a cyclohexane ring bridged by a methylene group in the 1,4-position, and is a bridged bicyclic compound. The compound can be synthesized by hydrogenation of the related compounds norbornene and norbornadiene.
Carbon	Carbon is the chemical element with symbol C and atomic number 6. As a member of group 14 on the periodic table, it is nonmetallic and tetravalent--making four electrons available to form covalent chemical bonds. There are three naturally occurring isotopes, with ^{12}C and ^{13}C being stable, while ^{14}C is radioactive, decaying with a half-life of about 5730 years.
1,2-dichloroethene	1,2-Dichloroethene, commonly called 1,2-dichloroethylene or 1,2-DCE, is an organochloride with the molecular formula $C_2H_2Cl_2$. It is a highly flammable, colorless liquid with a sharp, harsh odor. It can exist as either of two geometric isomers, cis-1,2-Dichloroethene or trans-1,2-Dichloroethene, but is often used as a mixture of the two.
Dehydrohalogenation	Dehydrohalogenation is an organic reaction from which an alkene is obtained from an alkyl halide . It is also called a β-Elimination reaction and is a type of elimination reaction. Ethanolic potassium hydroxide when reacted with alkyl halide gives alkene.
Dichlorocarbene	Dichlorocarbene is a carbene commonly encountered in organic chemistry. This reactive intermediate with chemical formula CCl_2 is easily available by reaction of chloroform and a base such as potassium t-butoxide or sodium hydroxide dissolved in water. A phase transfer catalyst for instance benzyltriethylammonium bromide is added to facilitate the migration of the hydroxide in the organic phase.
Alkyl	In chemistry, an Alkyl group is a hydrocarbon; typically an Alkyl is a part of a larger molecule. The term is usually used loosely, there is no general formula for an Alkyl group. In structural formulae, an Alkyl group is represented with an R. Usually, Alkyl groups resemble hydrocarbons, but with one less hydrogen atom.
Butylated hydroxyanisole	Butylated hydroxyanisole is an antioxidant consisting of a mixture of two isomeric organic compounds, 2-tert-butyl-4-hydroxyanisole and 3-tert-butyl-4-hydroxyanisole. It is prepared from 4-methoxyphenol and isobutylene. It is a waxy solid used in certain amounts as a food additive with the E number E320.
Diisopropylamine	Diisopropylamine is a secondary amine with the chemical formula $(CH_3)_2HC\text{-}NH\text{-}CH(CH_3)_2$. It is best known as its lithium salt, lithium diisopropylamide, known as 'LDA'. LDA is a strong, non-nucleophilic base.

Triethylamine	Triethylamine is the chemical compound with the formula $N(CH_2CH_3)_3$, commonly abbreviated Et_3N. It is also abbreviated TEA, yet this abbreviation must be used carefully to avoid confusion with triethanolamine, for which TEA is also a common abbreviation. It is commonly encountered in organic synthesis probably because it is the simplest symmetrically trisubstituted amine, i.e. a tertiary amine, that is liquid at room temperature. It possesses a strong fishy odor reminiscent of ammonia.
Phosphate	A Phosphate, an inorganic chemical, is a salt of phosphoric acid. Inorganic phosphates are mined to obtain phosphorus for use in agriculture and industry. In organic chemistry, a Phosphate, or organophosphate, is an ester of phosphoric acid.
Absolute configuration	An Absolute configuration in stereochemistry is the spatial arrangement of the atoms of a chiral molecular entity (or group) and its stereochemical description e.g. R or S. Absolute configuration s for chiral molecules are traditionally obtained by X-ray crystallography but only when the compound crystallises in one of the 65 Sohncke Groups (Chiral Space Groups.) Alternative techniques are Optical rotatory dispersion, vibrational circular dichroism and the use of chiral shift reagents in proton NMR. In step two the assignment of R or S is based on the Cahn-Ingold-Prelog priority rules. Absolute configuration s are also relevant to characterization of crystals.
Cyclohexane	Cyclohexane is a cycloalkane with the molecular formula C_6H_{12}. Cyclohexane is used as a nonpolar solvent for the chemical industry, and also as a raw material for the industrial production of adipic acid and caprolactam, both of which are intermediates used in the production of nylon. On an industrial scale, Cyclohexane is produced by reacting benzene with hydrogen.
Vicinal	In chemistry Vicinal stands for any two functional groups bonded to two adjacent carbon atoms. For example the molecule 2,3-dibromobutane carries two Vicinal bromine atoms and 1,3-dibromobutane does not. Likewise in a gem-dibromide the prefix gem, an abbreviation of geminal, signals that both bromine atoms are bonded to the same atom.
Zinc	Zinc is a metallic chemical element with the symbol Zn and atomic number 30. It is a first-row transition metal in group 12 of the periodic table. Zinc is chemically similar to magnesium because its ion is of similar size and its only common oxidation state is +2.
2-butanol	2-Butanol, or sec-butanol, is a organic compound with formula $CH_3CH(OH)CH_2CH_3$. This secondary alcohol is a flammable, colorless liquid that is soluble in 12 parts water and completely miscible with polar organic solvent such as ethers and other alcohols. It is produced on a large scale, primarily as a precursor to the industrial solvent methyl ethyl ketone.
Butanol	Butanol or butyl alcohol (sometimes also called bio Butanol when produced biologically), is a primary alcohol with a 4 carbon structure and the molecular formula of C_4H_9OH. It belongs to the higher alcohols and branched-chain alcohols. It is primarily used as a solvent, as an intermediate in chemical synthesis, and as a fuel.

	There are four isomeric structures for Butanol
Cycloalkanes	Cycloalkanes are types of alkanes which have one or more rings of carbon atoms in the chemical structure of their molecules. Alkanes are types of organic hydrocarbon compounds which have only single chemical bonds in their chemical structure. Cycloalkanes consist of only carbon and hydrogen atoms and are saturated because there are no multiple C-C bonds to hydrogenate
Dehydrogenation	Dehydrogenation is a chemical reaction that involves the elimination of hydrogen (H_2.) It is the reverse process of hydrogenation. Dehydrogenation reactions may be either large scale industrial processes or smaller scale laboratory procedures.
Petroleum	Petroleum or crude oil is a naturally occurring, flammable liquid found in rock formations in the Earth consisting of a complex mixture of hydrocarbons of various molecular weights, plus other organic compounds. The term 'Petroleum' was first used in the treatise De Natura Fossilium, published in 1546 by the German mineralogist Georg Bauer, also known as Georgius Agricola. The proportion of hydrocarbons in the mixture is highly variable and ranges from as much as 97% by weight in the lighter oils to as little as 50% in the heavier oils and bitumens.
Hofmann elimination	Hofmann elimination is a process where an amine is reacted to create a tertiary amine and an alkene by treatment with excess methyl iodide followed by treatment with silver oxide, water, and heat . After the first step, a quaternary ammonium iodide salt is created. After replacement of iodine by an hydroxyl anion, an elimination reaction takes place to the alkene.
Alkynes	Alkynes are hydrocarbons that have a triple bond between two carbon atoms, with the formula C_nH_{2n-2}. Alkynes are traditionally known as acetylenes, although the name acetylene also refers specifically to C_2H_2, known formally (but rarely) as ethyne using IUPAC nomenclature. Like other hydrocarbons, Alkynes are generally hydrophobic but tend to be more reactive.
Wittig reaction	The Wittig reaction is a chemical reaction of an aldehyde or ketone with a triphenyl phosphonium ylide (often called a Wittig reagent) to give an alkene and triphenylphosphine oxide. The Wittig reaction was discovered in 1954 by Georg Wittig, for which he was awarded the Nobel Prize in Chemistry in 1979. It is widely used in organic synthesis for the preparation of alkenes.

Alkene	In organic chemistry, an Alkene olefin, or olefine is an unsaturated chemical compound containing at least one carbon-to-carbon double bond. The simplest acyclic Alkene s, with only one double bond and no other functional groups, form a homologous series of hydrocarbons with the general formula C_nH_{2n}. The simplest Alkene is ethylene (C_2H_4), which has the International Union of Pure and Applied Chemistry (IUPAC) name ethene.
Carbon-carbon bond	A Carbon-carbon bond is a covalent bond between two carbon atoms. The most common form is the single bond - a bond composed of two electrons, one from each of the two atoms. The carbon-carbon single bond is a sigma bond and is said to be formed between one hybridized orbital from each of the carbon atoms.
Cracking	In petroleum geology and chemistry, Cracking is the process whereby complex organic molecules such as kerogens or heavy hydrocarbons are broken down into simpler molecules (e.g. light hydrocarbons) by the breaking of carbon-carbon bonds in the precursors. The rate of Cracking and the end products are strongly dependent on the temperature and presence of any catalysts. Cracking, also referred to as pyrolysis, is the breakdown of a large alkane into smaller, more useful alkanes and an alkene.
Hydrogenation	Hydrogenation is the chemical reaction that results from the addition of hydrogen (H_2.) The process is usually employed to reduce or saturate organic compounds. The process typically constitutes the addition of pairs of hydrogen atoms to a molecule.
Carbocation	A Carbocation is an ion with a positively-charged carbon atom. The charged carbon atom in a Carbocation is a 'sextet', i.e. it has only six electrons in its outer valence shell instead of the eight valence electrons that ensures maximum stability . Therefore Carbocation s are often reactive, seeking to fill the octet of valence electrons as well as regain a neutral charge.
Electrophilic addition	In organic chemistry, an Electrophilic addition reaction is an addition reaction where, in a chemical compound, a pi bond is removed by the creation of two new covalent bonds. The substrate of an Electrophilic addition reaction must have a double bond or triple bond . The driving force for this reaction is the formation of an electrophile X^+ that forms a covalent bond with an electron-rich unsaturated C=C bond.
Nucleophile	In chemistry, a Nucleophile is a reagent that forms a chemical bond to its reaction partner (the electrophile) by donating both bonding electrons. Because nucleophiles donate electrons, they are by definition Lewis bases All molecules or ions with a free pair of electrons can act as nucleophiles.
Hydrogen	Hydrogen is the chemical element with atomic number 1. It is represented by the symbol H. At standard temperature and pressure, Hydrogen is a colorless, odorless, nonmetallic, tasteless, highly flammable diatomic gas with the molecular formula H_2. With an atomic weight of 1.007 94 u, Hydrogen is the lightest element.
Alkynes	Alkynes are hydrocarbons that have a triple bond between two carbon atoms, with the formula C_nH_{2n-2}. Alkynes are traditionally known as acetylenes, although the name acetylene also refers specifically to C_2H_2, known formally (but rarely) as ethyne using IUPAC nomenclature. Like other hydrocarbons, Alkynes are generally hydrophobic but tend to be more reactive.

Halide	A Halide is a binary compound, of which one part is a halogen atom and the other part is an element or radical that is less electronegative than the halogen, to make a fluoride, chloride, bromide, iodide, or astatide compound. Many salts are Halide s. All Group 1 metals form Halide s with the halogens and they are white solids.
Alcohol	In chemistry, an Alcohol is any organic compound in which a hydroxyl group (-OH) is bound to a carbon atom of an alkyl or substituted alkyl group. The general formula for a simple acyclic Alcohol is $C_nH_{2n+1}OH$. In common terms, the word Alcohol refers to ethanol, the type of Alcohol found in Alcohol ic beverages. Ethanol is a colorless, volatile liquid with a mild odor which can be obtained by the fermentation of sugars.
Propene	Propene, also known as propylene, is an unsaturated organic compound having the chemical formula C_3H_6. It has one double bond, and is the second simplest member of the alkene class of hydrocarbons, and it is also second in natural abundance. At room temperature, Propene is a colourless, odourless gas, though when used as a fuel, it is mixed with minute quantities of foul-smelling sulfurous compounds (mercaptans) so that gas leaks can be readily detected.
2,2-dimethylbutane	2,2-Dimethylbutane, trivially known as neohexane, is the isomer of hexane containing a quaternary carbon.It is a hydrocarbon i.e it contains hydrogen and carbon only. It has covalent bonds and is single bonded.
Dimethylamine	Dimethylamine is an organic compound with the formula $(CH_3)_2NH$. This secondary amine is a colorless, flammable liquified gas with an ammonia- or fish-like odor. Dimethylamine is generally encountered as a solution in water at concentrations up to around 40%. In 2005, an estimated 270,000 tons were produced.
Toluene	Toluene phenylmethane, and Toluol, is a clear water-insoluble liquid with the typical smell of paint thinners, redolent of the sweet smell of the related compound benzene. It is an aromatic hydrocarbon that is widely used as an industrial feedstock and as a solvent. Like other solvents, Toluene is also used as an inhalant drug for its intoxicating properties; however this causes severe neurological harm.
Diborane	Diborane is the chemical compound consisting of boron and hydrogen with the formula B_2H_6. It is a colorless gas at room temperature with a repulsively sweet odor. Diborane mixes well with air, easily forming explosive mixtures.
Tetrahydrofuran	Tetrahydrofuran is a colorless, water-miscible organic liquid with low-viscosity at standard temperature and pressure. It is a heterocyclic compound with a chemical formula C_4H_8O, and is the fully hydrogenated analog of the aromatic organic compound furan. It is one of the most polar of the organic functional class of ethers, and has a relatively low freeze point, and so is a commonly used modern organic chemical laboratory solvent across a range of temperatures.
Isomer	In chemistry, Isomer s are compounds with the same molecular formula but different structural formula. Isomer s do not necessarily share similar properties unless they also have the same functional groups. This should not be confused with a nuclear Isomer which involves a nucleus at different states of excitement.
Norbornene	Norbornene or norbornylene or norcamphene is a bridged cyclic hydrocarbon. It is a white solid with a pungent sour odor. The molecule consists of a cyclohexene ring bridged with a methylene group in the para position.
Halogenation	Halogenation is a chemical reaction that incorporates a halogen atom into a molecule. More specific descriptions exist that specify the type of halogen: fluorination, chlorination, bromination, and iodination.

In a Markovnikov addition reaction, a halogen like bromine is reacted with an alkene which causes the π-bond to break forming an haloalkane.

Vicinal

In chemistry Vicinal stands for any two functional groups bonded to two adjacent carbon atoms. For example the molecule 2,3-dibromobutane carries two Vicinal bromine atoms and 1,3-dibromobutane does not.

Likewise in a gem-dibromide the prefix gem, an abbreviation of geminal, signals that both bromine atoms are bonded to the same atom.

Alkanes

Alkanes are chemical compounds that consist only of the elements carbon and hydrogen, wherein these atoms are linked together exclusively by single bonds without any cyclic structure Alkanes belong to a homologous series of organic compounds in which the members differ by a constant relative atomic mass of 14.

Each carbon atom must have 4 bonds, and each hydrogen atom must be joined to a carbon atom

Alkyl

In chemistry, an Alkyl group is a hydrocarbon; typically an Alkyl is a part of a larger molecule. The term is usually used loosely, there is no general formula for an Alkyl group. In structural formulae, an Alkyl group is represented with an R. Usually, Alkyl groups resemble hydrocarbons, but with one less hydrogen atom.

Aryl

In the context of organic molecules, Aryl refers to any functional group or substituent derived from a simple aromatic ring, may it be phenyl, thiophenyl, indolyl, etc . 'Aryl' is used for the sake of abbreviation or generalization.

A simple Aryl group is phenyl, C_6H_5; it is derived from benzene.

Cyclohexane

Cyclohexane is a cycloalkane with the molecular formula C_6H_{12}. Cyclohexane is used as a nonpolar solvent for the chemical industry, and also as a raw material for the industrial production of adipic acid and caprolactam, both of which are intermediates used in the production of nylon. On an industrial scale, Cyclohexane is produced by reacting benzene with hydrogen.

Cyclohexene

Cyclohexene is a colorless clear liquid cycloalkene with an intense aversive characteristic sharp smell reminiscent of an oil refinery.

It is not very stable upon long term storage with exposure to light and air and should be distilled before use to eliminate peroxides. A common experiment among beginning organic chemistry students is the acid catalyzed dehydration of cyclohexanol with distillative removal of the resulting Cyclohexene from the reaction mixture:

- Refractive Index: 1.4465
- Critical temperature: 287.2 °C (560.4 K)

Cyclopentene

Cyclopentene is a chemical compound with the formula C_5H_8. It is a colorless liquid with a petrol-like odor. It is one of the cycloalkenes.

Halohydrin

A Halohydrin is a type of chemical compound or functional group in which one carbon atom has a halogen substituent, and an adjacent carbon atom has a hydroxyl substituent.

Halohydrin formation:

- from an alkene in a Halohydrin formation reaction
- from an epoxide by a hydrohalic acid

Halohydrin reactions:

- In presence of a base, such as potassium hydroxide, a Halohydrin may undergo internal S_N2 reaction to form an epoxide. This is the reverse of the formation reaction from an epoxide.
- Epoxidation in biological systems can be catalyzed by Halohydrin dehalogenase.

.

Asymmetric induction

Asymmetric induction in stereochemistry describes the preferential formation in a chemical reaction of one enantiomer or diastereoisomer over the other as a result of the influence of a chiral feature present in the substrate, reagent, catalyst or environment. Asymmetric induction is a key element in asymmetric synthesis.

Asymmetric induction was introduced by Emil Fischer based on his work on carbohydrates.

Carbene

In chemistry, a Carbene is a organic molecule containing a carbon atom with six valence electrons and having the general formula RR'C:. Carbene s are classified into two varieties, singlets and triplets. Most Carbene s are very short lived, although persistent Carbene s are known.

Cyclopropane

Cyclopropane is a cycloalkane molecule with the molecular formula C_3H_6, consisting of three carbon atoms linked to each other to form a ring, with each carbon atom bearing two hydrogen atoms.

The bonds between the carbon atoms are considerably weaker than in a typical carbon-carbon bond, yielding reactivity similar to or greater than alkenes. Baeyer strain theory explains why: the angle strain from the 60° angle between the carbon atoms (less than the normal angle of 109.5° for bonds between atoms with sp^3 hybridised orbitals) reduces the compound's carbon-carbon bond energy, making it more reactive than other cycloalkanes such as cyclohexane and cyclopentane.

Diazomethane

Diazomethane is the chemical compound CH_2N_2. It is one of the more common diazo compounds. In the pure form at room temperature, it is a famously explosive yellow gas, but it is almost universally used as a solution in diethyl ether.

Isoniazid

Isoniazid is an organic compound that is the first-line antituberculosis medication in prevention and treatment. First discovered in 1912 as an inhibitor of the MAO enzyme, it was first used as an antidepressant, but discontinued due to side effects. In 1951, it was later discovered that Isoniazid was effective against TB. Isoniazid is never used on its own to treat active tuberculosis because resistance quickly develops.

Methylene	Methylene is a chemical species in which a carbon atom is bonded to two hydrogen atoms. Three different possibilities present themselves: • the $-CH_2-$ group, e.g. dichloromethane (also known as Methylene chloride) • the $=CH_2$ group, e.g. methylenecyclopropene, • and the $:CH_2$ molecule, a carbene known as Methylene. A carbene is a highly reactive organic molecule having a divalent carbon atom with six valence electrons. Methylene groups in a chain or ring contribute to its size and lipophilicity.
Carbenoid	In chemistry a Carbenoid is a reactive intermediate that shares reaction characteristics with a carbene . In the Simmons-Smith reaction the Carbenoid intermediate is a zinc / iodine complex that takes the form of $I-CH_2-Zn-I$ This complex reacts with an alkene to form a cyclopropane just as a carbene would do. Carbenoid s appear as intermediates in many other reactions.
Simmons-Smith reaction	The Simmons-Smith reaction is an organic reaction in which a carbenoid reacts with an alkene (or alkyne) to form a cyclopropane. It is named after Howard Ensign Simmons, Jr. and R. D. Smith.
Bromoform	Bromoform is a pale yellowish liquid with a sweet odor similar to chloroform, a halomethane or haloform. Its refractive index is 1.595 (20 °C, D.) Small amounts are formed naturally by plants in the ocean.
Dehydrohalogenation	Dehydrohalogenation is an organic reaction from which an alkene is obtained from an alkyl halide . It is also called a β-Elimination reaction and is a type of elimination reaction. Ethanolic potassium hydroxide when reacted with alkyl halide gives alkene.
Alpha carbon	The alpha carbon in organic chemistry refers to the first carbon that attaches to a functional group (the carbon is attached at the first position.) By extension, the second carbon is the beta carbon, and so on. This nomenclature can also be applied to the hydrogen atoms attached to the carbons.
Acetic acid	Acetic acid, CH_3COOH, also known as ethanoic acid, is an organic acid which gives vinegar its sour taste and pungent smell. Pure, water-free Acetic acid is a colourless liquid that absorbs water from the environment (hygroscopy), and freezes at 16.7 °C (62 °F) to a colourless crystalline solid. It is a weak acid, in that it is only partially dissociated acid in aqueous solution.
Benzoic acid	Benzoic acid, $C_7H_6O_2$ (or C_6H_5COOH), is a colorless crystalline solid and the simplest aromatic carboxylic acid. The name derived from gum benzoin, which was for a long time the only source for Benzoic acid. This weak acid and its salts are used as a food preservative.
Carboxylic acids	Carboxylic acids are organic acids characterized by the presence of a carboxyl group, which has the formula -C(=O)OH, usually written -COOH or $-CO_2H$. Carboxylic acids are Brønsted-Lowry acids -- they are proton donors. Salts and anions of Carboxylic acids are called carboxylates.

The simplest series of Carboxylic acids are the alkanoic acids, R-COOH, where R is a hydrogen or an alkyl group.

Epoxide

An Epoxide is a cyclic ether with three ring atoms. This ring approximately defines an equilateral triangle, which makes it highly strained. The strained ring makes Epoxide s more reactive than other ethers.

Peroxybenzoic acid

Peroxybenzoic acid is a simple peroxy acid. It may be synthesized from benzoic acid and hydrogen peroxide, or by the treatment of benzoyl peroxide with sodium methoxide, followed by acidification.

Like other peroxyacids, it may be used to generate epoxides, such as styrene oxide from styrene:

.

Amide

In chemistry, an Amide is one of three kinds of compounds:

- (sometimes called acid Amide the organic functional group characterized by a carbonyl group (C=O) linked to a nitrogen atom (N), or a compound that contains this functional group (pictured to the right); or
- a kind of anion or
- any organic compound derived by the replacement of a hydroxyl group by an amino group.

Amide s are the most stable of all the carbonyl functional groups.

Many chemists make a pronunciation distinction between the two, saying /É™ËˆmiË□d/ for the carbonyl-nitrogen compound and /ËˆæmaÉªd/ for the anion. Others substitute one of these with , while still others pronounce both , making them homonyms.

In the first sense referred to above, an Amide is an amine where one of the nitrogen substituents is an acyl group; it is generally represented by the formula: $R_1(CO)NR_2R_3$, where either or both R_2 and R_3 may be hydrogen.

Reagent

A Reagent or reactant is a substance or compound consumed during a chemical reaction. Solvents and catalysts, although they are involved in the reaction, are usually not referred to as reactants.

Although the terms reactant and Reagent are often used interchangeably, a Reagent is more specifically 'a test substance that is added to a system in order to bring about a reaction or to see whether a reaction occurs'.

Osmium tetroxide

Osmium tetroxide is the chemical compound with the formula OsO_4. The compound is noteworthy for its many uses, despite the rarity of osmium. It also has a number of interesting properties, one being that the solid is volatile.

Manganese dioxide

Manganese dioxide is the inorganic compound with the formula MnO_2. This blackish or brown solid occurs naturally as the mineral pyrolusite, which is the main ore of manganese. It is also present in manganese nodules.

Potassium permanganate	Potassium permanganate is the inorganic chemical compound $KMnO_4$, a water soluble salt consisting of equal mole amounts of potassium (K^+) and permanganate (MnO_4^-, officially called manganate (VII)) ions. This salt, formerly known as permanganate of potash or Condy's crystals is a strong oxidizing agent. It dissolves in water to give deep purple solutions, evaporation of which gives prismatic purplish-black glistening crystals.
Ozone	Ozone or trioxygen (O_3) is a triatomic molecule, consisting of three oxygen atoms. It is an allotrope of oxygen that is much less stable than the diatomic O_2. Ground-level Ozone is an air pollutant with harmful effects on the respiratory systems of animals.
Ozonolysis	Ozonolysis is the cleavage of an alkene or alkyne with ozone to form organic compounds in which the multiple carbon-carbon bond has been replaced by a double bond to oxygen. The outcome of the reaction depends on the type of multiple bond being oxidized and the workup conditions. Alkenes can be oxidized with ozone to form alcohols, aldehydes or ketones, or carboxylic acids.
Ether	Ether is a class of organic compounds which contain an Ether group -- an oxygen atom connected to two (substituted) alkyl or aryl groups -- of general formula R-O-R'. A typical example is the solvent and anesthetic diethyl Ether, commonly referred to simply as 'Ether' (ethoxyethane, CH_3-CH_2-O-CH_2-CH_3.) Ether molecules cannot form hydrogen bonds amongst each other, resulting in a relatively low boiling point compared to that of the analogous alcohols.
Molozonide	A Molozonide is a term for a 1,2,3-trioxolane. It is an unstable cyclic organic compound, which includes three oxygen atoms and two sp^3-hybridized carbon atoms. A Molozonide is an intermediate during ozonolysis as a precursor of the ozonide complex formed during cleavage.
Isobutylene	Isobutylene is a hydrocarbon of significant industrial importance. It is a four-carbon branched alkene (olefin), one of the four isomers of butylene. At standard temperature and pressure it is a colorless flammable gas.
Styrene	Styrene, also known as vinyl benzene as well as many other names , is an organic compound with the chemical formula $C_6H_5CH=CH_2$. This aromatic hydrocarbon is a colorless oily liquid that evaporates easily and has a sweet smell, although high concentrations confer a less pleasant odor. Styrene is the precursor to polystyrene and several copolymers.
Ethylene	Ethylene is the chemical compound with the formula C_2H_4. It is the simplest alkene. Because it contains a carbon-carbon double bond, Ethylene is called an unsaturated hydrocarbon or an olefin.
Polyethylene	Polyethylene or polythene (IUPAC name polyethene or poly(methylene)) is a thermoplastic commodity heavily used in consumer products (notably the plastic shopping bag.) Over 60 million tons of the material are produced worldwide every year. Polyethylene is a polymer consisting of long chains of the monomer ethylene (IUPAC name ethene.)

Organic synthesis	Organic synthesis is a special branch of chemical synthesis and is concerned with the construction of organic compounds via organic reactions. Organic molecules can often contain a higher level of complexity compared to purely inorganic compounds, so the synthesis of organic compounds has developed into one of the most important branches of organic chemistry. There are two main areas of research fields within the general area of Organic synthesis: total synthesis and methodology.
Retrosynthetic analysis	Retrosynthetic analysis is a technique for solving problems in the planning of organic syntheses. This is achieved by transforming a target molecule into simpler precursor structures without assumptions regarding starting materials. Each precursor material is examined using the same method.

Carbon-carbon bond	A Carbon-carbon bond is a covalent bond between two carbon atoms. The most common form is the single bond - a bond composed of two electrons, one from each of the two atoms. The carbon-carbon single bond is a sigma bond and is said to be formed between one hybridized orbital from each of the carbon atoms.
Cicutoxin	Cicutoxin is a poisonous polyyne and alcohol found in various plants, most notably water hemlock (Cicuta species.) It is structurally related to the oenanthotoxin of hemlock water dropwort. It is a cholinergic poison, causing death by disruption of the central nervous system.
Dimethylacetylene	Dimethylacetylene is an alkyne with chemical formula $CH_3C{\equiv}CCH_3$. Produced artificially, it is a colorless, volatile, pungent liquid at standard temperature and pressure.
Ethylacetylene	Ethylacetylene, also known as 1-butyne, but-1-yne, ethylethyne, and UN 2452, is an extremely flammable and reactive alkyne with chemical formula C_4H_6 and CAS number 107-00-6 that is used in the synthesis of organic compounds. It occurs as a colorless gas.
Ethylene	Ethylene is the chemical compound with the formula C_2H_4. It is the simplest alkene. Because it contains a carbon-carbon double bond, Ethylene is called an unsaturated hydrocarbon or an olefin.
Estradiol	Estradiol (E2 or 17β Estradiol (also o Estradiol) is a sex hormone. Mislabelled the 'female' hormone, it is also present in males; it represents the major estrogen in humans. Estradiol has not only a critical impact on reproductive and sexual functioning, but also affects other organs including the bones.
Acetylene	Acetylene is the chemical compound with the formula HC_2H. It is a hydrocarbon and the simplest alkyne. This colourless gas is widely used as a fuel and a chemical building block. It is unstable in pure form and thus is usually handled as a solution.
Alkynes	Alkynes are hydrocarbons that have a triple bond between two carbon atoms, with the formula C_nH_{2n-2}. Alkynes are traditionally known as acetylenes, although the name acetylene also refers specifically to C_2H_2, known formally (but rarely) as ethyne using IUPAC nomenclature. Like other hydrocarbons, Alkynes are generally hydrophobic but tend to be more reactive.
Diphenylacetylene	Diphenylacetylene is the chemical compound $C_6H_5C{\equiv}CC_6H_5$. The molecule consists of phenyl groups attached to both ends of an alkyne. It is a colorless crystalline material that is widely used as a building block in organic and as a ligand in organometallic chemistry.
Methoxyethane	Methoxyethane is a colorless gaseous ether with a medicine-like odor. It is extremely flammable, and its inhalation may cause asphyxiation or dizzyness. As a Lewis base, it can react with Lewis acids to form salts and reacts violently with oxidizing agents.
Methylacetylene	Methylacetylene is an alkyne with the chemical formula $CH_3C{\equiv}CH$. It is a component of MAPP gas along with its isomer 1,2-propadiene (allene), which is commonly used in gas welding. Unlike acetylene, Methylacetylene can be safely condensed.

Methylacetylene exists in equilibrium with allene, the mixture of Methylacetylene and propadiene being called MAPD:

$H_3CC{\equiv}CH$ □ $H_2C{=}C{=}CH_2$

K_{eq} = 0.22 (270 °C), 0.1 K (5 °C) MAPD is produced as a side product, often an undesirable one, by cracking propane to produce propene, an important feedstock in the chemical industry.

Phenylacetylene

Phenylacetylene is an alkyne hydrocarbon containing a phenyl group. It exists as a colorless, viscous liquid. In research, it is sometimes used as an analog for acetylene; being a liquid, it is easier to handle than acetylene gas.

Propargyl

In organic chemistry, Propargyl is an alkyl functional group of 2-propynyl with the structure $HC{\equiv}C{-}CH_2{-}$, derived from the alkyne propyne.

The term propargylic refers to a saturated position (sp^3-hybridized) on a molecular framework next to an alkynyl group;

The term homopropargylic designates in the same manner

- a saturated position on a molecular framework next to a propargylic group and thus two bonds from an alkyne moiety;
- a 3-butynyl fragment, $HC{\equiv}C{-}CH_2CH_2{-}$, or substituted homologue.

.

Propargyl alcohol

Propargyl alcohol is an organic compound which is a simple alcohol containing an alkyne functional group . Propargyl alcohol is a clear colorless viscous liquid that is miscible with water and most polar organic solvents. It is insoluble in most hydrocarbon solvents.

Allene

An Allene is a hydrocarbon in which one atom of carbon is connected by double bonds with two other atoms of carbon. Allene also is the common name for the parent compound of this series, propadiene.

Such pair of bonds make Allene s much more reactive than other alkenes.

Linear

In chemistry, the Linear molecular geometry describes the arrangement of three or more atoms placed at an expected bond angle of 180Â°. Linear organic molecules, e.g. acetylene, are often described by invoking sp orbital hybridization for the carbon centers. Many Linear molecules exist, prominent examples include CO_2, HCN, and xenon difluoride.

Sodium

Sodium is a metallic element with a symbol Na and atomic number 11. It is a soft, silvery-white, highly reactive metal and is a member of the alkali metals within 'group 1' (formerly known as 'group IA'.) It has only one stable isotope, ^{23}Na.

Sodium amide

Sodium amide, commonly called sodamide, is the chemical compound with the formula $NaNH_2$. This solid, which is dangerously reactive toward water, is white when pure, but commercial samples are typically gray due to the presence of small quantities of metallic iron from the manufacturing process. Such impurities do not usually affect the utility of the reagent.

1-hexene	1-Hexene is a higher olefin, or alkene, with a formula C_6H_{12}. 1-Hexene is an alpha-olefin, meaning that the double bond is located at the alpha (primary) position, endowing the compound with higher reactivity and thus useful chemical properties. 1-Hexene is an industrially significant linear alpha olefin.
Alkylation	Alkylation is the transfer of an alkyl group from one molecule to another. The alkyl group may be transferred as an alkyl carbocation, a free radical, a carbanion or a carbene (or their equivalents) . Alkylating agents are widely used in chemistry because the alkyl group is probably the most common group encountered in organic molecules.
Ether	Ether is a class of organic compounds which contain an Ether group -- an oxygen atom connected to two (substituted) alkyl or aryl groups -- of general formula R-O-R'. A typical example is the solvent and anesthetic diethyl Ether, commonly referred to simply as 'Ether' (ethoxyethane, CH_3-CH_2-O-CH_2-CH_3.) Ether molecules cannot form hydrogen bonds amongst each other, resulting in a relatively low boiling point compared to that of the analogous alcohols.
Isomer	In chemistry, Isomer s are compounds with the same molecular formula but different structural formula. Isomer s do not necessarily share similar properties unless they also have the same functional groups. This should not be confused with a nuclear Isomer which involves a nucleus at different states of excitement.
Formaldehyde	Formaldehyde (IUPAC name methanal) is a chemical compound with the formula CH_2O. It is the simplest aldehyde. Formaldehyde also exists as the cyclic trimer trioxane and the polymer para Formaldehyde . It exists in water as the hydrate $H_2C(OH)_2$.
Ketone	In organic chemistry, a Ketone is a type of compound that features a carbonyl group bonded to two other carbon atoms, i.e., R_3CCO-CR_3 where R can be a variety of atoms and groups of atoms. With carbonyl carbon bonded to two carbon atoms, Ketone s are distinct from many other functional groups, such as carboxylic acids, aldehydes, esters, amides, and other oxygen-containing compounds. The double-bond of the carbonyl group distinguishes Ketone s from alcohols and ethers.
Dehydrohalogenation	Dehydrohalogenation is an organic reaction from which an alkene is obtained from an alkyl halide . It is also called a β-Elimination reaction and is a type of elimination reaction. Ethanolic potassium hydroxide when reacted with alkyl halide gives alkene.
Halide	A Halide is a binary compound, of which one part is a halogen atom and the other part is an element or radical that is less electronegative than the halogen, to make a fluoride, chloride, bromide, iodide, or astatide compound. Many salts are Halide s. All Group 1 metals form Halide s with the halogens and they are white solids.
Alkyl	In chemistry, an Alkyl group is a hydrocarbon; typically an Alkyl is a part of a larger molecule. The term is usually used loosely, there is no general formula for an Alkyl group. In structural formulae, an Alkyl group is represented with an R. Usually, Alkyl groups resemble hydrocarbons, but with one less hydrogen atom.
Aryl	In the context of organic molecules, Aryl refers to any functional group or substituent derived from a simple aromatic ring, may it be phenyl, thiophenyl, indolyl, etc . 'Aryl' is used for the sake of abbreviation or generalization. A simple Aryl group is phenyl, C_6H_5; it is derived from benzene.

Geminal

In chemistry, the term Geminal refers to the relationship between two functional groups that are attached to the same atom. The prefix gem is applied to a chemical name to denote this relationship, as in a gem-dibromide.

The following example shows the conversion of a cyclohexyl methyl ketone to a gem-dichloride through a reaction with phosphorus pentachloride.

Vicinal

In chemistry Vicinal stands for any two functional groups bonded to two adjacent carbon atoms. For example the molecule 2,3-dibromobutane carries two Vicinal bromine atoms and 1,3-dibromobutane does not.

Likewise in a gem-dibromide the prefix gem, an abbreviation of geminal, signals that both bromine atoms are bonded to the same atom.

Addition reaction

An Addition reaction in organic chemistry, is in its simplest terms an organic reaction where two or more molecules combine to form a larger one.

There are two main types of polar Addition reaction s:

- Electrophilic addition
- Nucleophilic addition

Other non-polar Addition reaction s exists as well:

- Free radical addition

Addition reaction s are limited to chemical compounds that have multiply-bonded atoms:

- Molecules with carbon-carbon double bonds (alkenes) or triple bonds (alkynes)
- Molecules with carbon - hetero double bonds like C=O or C=N

An Addition reaction is the opposite of an elimination reaction. For instance the hydration reaction of an alkene and the dehydration of an alcohol are addition-elimination pairs.

In the related Addition-elimination reaction an Addition reaction is followed by an elimination reaction.

Butane

Butane also called n Butane is the unbranched alkane with four carbon atoms, $CH_3CH_2CH_2CH_3$. Butane is also used as a collective term for n Butane together with its only other isomer, iso Butane (also called methylpropane), $CH(CH_3)_3$.

Butane s are highly flammable, colorless, odorless, easily liquefied gases.

Cracking

In petroleum geology and chemistry, Cracking is the process whereby complex organic molecules such as kerogens or heavy hydrocarbons are broken down into simpler molecules (e.g. light hydrocarbons) by the breaking of carbon-carbon bonds in the precursors. The rate of Cracking and the end products are strongly dependent on the temperature and presence of any catalysts. Cracking, also referred to as pyrolysis, is the breakdown of a large alkane into smaller, more useful alkanes and an alkene.

Hydrogenation

Hydrogenation is the chemical reaction that results from the addition of hydrogen (H_2.) The process is usually employed to reduce or saturate organic compounds. The process typically constitutes the addition of pairs of hydrogen atoms to a molecule.

Hexene

Hexene is a higher olefin, or alkene with a formula C_6H_{12}. The 'Hex' is derived from the fact that there are 6 carbon atoms in the molecule, while the 'ene' suffix denotes that two carbon atoms are connected via a double bond. There are several isomers of Hexene, depending on the position of the double bond and the branching of the carbon chain.

Octene

Octene is an alkene, or higher olefin, with the formula C_8H_{16}. There are several isomers of Octene, depending on the position of the double bond and the branching of the carbon chain. One of the most common industrially useful isomers is 1-Octene, an alpha-olefin, used primarily as a co-monomer in production of polyethylene via the solution polymerization process.

Halogenation

Halogenation is a chemical reaction that incorporates a halogen atom into a molecule. More specific descriptions exist that specify the type of halogen: fluorination, chlorination, bromination, and iodination.

In a Markovnikov addition reaction, a halogen like bromine is reacted with an alkene which causes the π-bond to break forming an haloalkane.

Hydrogen

Hydrogen is the chemical element with atomic number 1. It is represented by the symbol H. At standard temperature and pressure, Hydrogen is a colorless, odorless, nonmetallic, tasteless, highly flammable diatomic gas with the molecular formula H_2. With an atomic weight of 1.007 94 u, Hydrogen is the lightest element.

Alkanes

Alkanes are chemical compounds that consist only of the elements carbon and hydrogen, wherein these atoms are linked together exclusively by single bonds without any cyclic structure Alkanes belong to a homologous series of organic compounds in which the members differ by a constant relative atomic mass of 14.

Each carbon atom must have 4 bonds, and each hydrogen atom must be joined to a carbon atom

Alkene

In organic chemistry, an Alkene olefin, or olefine is an unsaturated chemical compound containing at least one carbon-to-carbon double bond. The simplest acyclic Alkene s, with only one double bond and no other functional groups, form a homologous series of hydrocarbons with the general formula C_nH_{2n}.

The simplest Alkene is ethylene (C_2H_4), which has the International Union of Pure and Applied Chemistry (IUPAC) name ethene.

Enol

Enol s are alkenes with a hydroxyl group affixed to one of the carbon atoms composing the double bond. Enol s and carbonyl compounds are in fact isomers; this is called keto Enol tautomerism:

The Enol form is shown above. It is usually unstable, does not survive long, and changes into the keto form shown on the right.

Keto-enol tautomerism

In organic chemistry, Keto-enol tautomerism refers to a chemical equilibrium between a keto form (a ketone or an aldehyde) and an enol. The enol and keto forms are said to be tautomers of each other. The interconversion of the two forms involves the movement of a proton and the shifting of bonding electrons; hence, the isomerism qualifies as tautomerism.

Esterification

Esterification is the general name for a chemical reaction in which two reactants (typically an alcohol and an acid) form an ester as the reaction product. Esters are common in organic chemistry and biological materials, and often have a characteristic pleasant, fruity odor. This leads to their extensive use in the fragrance and flavor industry.

Pentyl

In organic chemistry, Pentyl is a five-carbon alkyl substituent with chemical formula $-C_5H_{11}$. It is the substituent form of the alkane pentane.

- Pentanol
- Pentyl pentanoate
- Amylamine
- Amyl acetate
- Amyl alcohol

See Amyl.

Potassium permanganate

Potassium permanganate is the inorganic chemical compound $KMnO_4$, a water soluble salt consisting of equal mole amounts of potassium (K^+) and permanganate (MnO_4^-, officially called manganate (VII)) ions. This salt, formerly known as permanganate of potash or Condy's crystals is a strong oxidizing agent. It dissolves in water to give deep purple solutions, evaporation of which gives prismatic purplish-black glistening crystals.

Ozonolysis

Ozonolysis is the cleavage of an alkene or alkyne with ozone to form organic compounds in which the multiple carbon-carbon bond has been replaced by a double bond to oxygen. The outcome of the reaction depends on the type of multiple bond being oxidized and the workup conditions.

Alkenes can be oxidized with ozone to form alcohols, aldehydes or ketones, or carboxylic acids.

Alcohol	In chemistry, an Alcohol is any organic compound in which a hydroxyl group (-OH) is bound to a carbon atom of an alkyl or substituted alkyl group. The general formula for a simple acyclic Alcohol is $C_nH_{2n+1}OH$. In common terms, the word Alcohol refers to ethanol, the type of Alcohol found in Alcohol ic beverages. Ethanol is a colorless, volatile liquid with a mild odor which can be obtained by the fermentation of sugars.
Carbinol	Methanol, also known as methyl alcohol, Carbinol, wood alcohol, wood naphtha or wood spirits, is a toxic chemical with chemical formula CH_3OH Drinking even small amounts can cause blindness. It is the simplest alcohol, and is a light, volatile, colourless, flammable, toxic liquid with a distinctive odor that is very similar to but slightly sweeter than ethanol (drinking alcohol.)
Carbon	Carbon is the chemical element with symbol C and atomic number 6. As a member of group 14 on the periodic table, it is nonmetallic and tetravalent--making four electrons available to form covalent chemical bonds. There are three naturally occurring isotopes, with ^{12}C and ^{13}C being stable, while ^{14}C is radioactive, decaying with a half-life of about 5730 years.
Ethanol	Ethanol pure alcohol, grain alcohol is a volatile, flammable, colorless liquid. It is a psychoactive drug, best known as the type of alcohol found in alcoholic beverages and in modern thermometers. Ethanol is one of the oldest recreational drugs.
Grignard reaction	The Grignard reaction is an organometallic chemical reaction in which alkyl- or aryl-magnesium halides , act as nucleophiles, attack electrophilic carbon atoms that are present within polar bonds to yield a carbon-carbon bond (compare to Wittig reaction), thus altering hybridization about the reaction center. The Grignard reaction is an important tool in the formation of carbon-carbon bonds and for the formation of carbon-phosphorus, carbon-tin, carbon-silicon, carbon-boron and other carbon-heteroatom bonds. The addition to the nucleophile is irreversible due to the high pK_a value of the alkyl component (pK_a = ~45.)
Methanol	Methanol, also known as methyl alcohol, carbinol, wood alcohol, wood naphtha or wood spirits, is a toxic chemical with chemical formula CH_3OH Drinking even small amounts can cause blindness. It is the simplest alcohol, and is a light, volatile, colourless, flammable, toxic liquid with a distinctive odor that is very similar to but slightly sweeter than ethanol (drinking alcohol.)
Alpha carbon	The alpha carbon in organic chemistry refers to the first carbon that attaches to a functional group (the carbon is attached at the first position.) By extension, the second carbon is the beta carbon, and so on. This nomenclature can also be applied to the hydrogen atoms attached to the carbons.
Phenol	In organic chemistry, phenols, sometimes called phenolics, are a class of chemical compounds consisting of a hydroxyl group (-OH) bonded directly to an aromatic hydrocarbon group. The simplest of the class is Phenol Phenol - the simplest of the phenols. Although similar to alcohols, phenols have unique properties and are not classified as alcohols (since the hydroxyl group is not bonded to a saturated carbon atom.)

Butanol	Butanol or butyl alcohol (sometimes also called bio Butanol when produced biologically), is a primary alcohol with a 4 carbon structure and the molecular formula of C_4H_9OH. It belongs to the higher alcohols and branched-chain alcohols. It is primarily used as a solvent, as an intermediate in chemical synthesis, and as a fuel. There are four isomeric structures for Butanol
Substituent	In organic chemistry and biochemistry, a Substituent is an atom or group of atoms substituted in place of a hydrogen atom on the parent chain of a hydrocarbon. The terms Substituent, side chain, group, branch, or pendant group are used almost interchangeably to describe branches from a parent structure, though certain distinctions are made in the context of polymer chemistry. In polymers, side chains extend from a backbone structure.
Alkene	In organic chemistry, an Alkene olefin, or olefine is an unsaturated chemical compound containing at least one carbon-to-carbon double bond. The simplest acyclic Alkene s, with only one double bond and no other functional groups, form a homologous series of hydrocarbons with the general formula C_nH_{2n}. The simplest Alkene is ethylene (C_2H_4), which has the International Union of Pure and Applied Chemistry (IUPAC) name ethene.
Cyclopentane	Cyclopentane is a highly flammable alicyclic hydrocarbon with chemical formula C_5H_{10} and CAS number 287-92-3, consisting of a ring of five carbon atoms each bonded with two hydrogen atoms above and below the plane. It occurs as a colorless liquid with a petrol-like odor. Its melting point is −94 °C and its boiling point is 49 °C. The typical structure of Cyclopentane is the 'envelope' conformation. Cyclopentane is used in the manufacture of synthetic resins and rubber adhesives and also as a blowing agent in the manufacture of polyurethane insulating foam, as found in may domestic appliances such as refrigerators and freezers, replacing environmentally damaging alternatives such as CFC-11 and HCFC-141b More advanced technologies, such as computer hard drives and outerspace equipment employ multiply-alkylated Cyclopentane lubricants because of their extremely low volatility.
Diol	A Diol or glycol is a chemical compound containing two hydroxyl groups (-OH groups) Vicinal Diol s have hydroxyl groups attached to adjacent atoms. Examples of vicinal Diol compounds are ethylene glycol and propylene glycol. Geminal Diol s have hydroxyl groups bonded to the same atom.
Vicinal	In chemistry Vicinal stands for any two functional groups bonded to two adjacent carbon atoms. For example the molecule 2,3-dibromobutane carries two Vicinal bromine atoms and 1,3-dibromobutane does not. Likewise in a gem-dibromide the prefix gem, an abbreviation of geminal, signals that both bromine atoms are bonded to the same atom.

Isomer	In chemistry, Isomer s are compounds with the same molecular formula but different structural formula. Isomer s do not necessarily share similar properties unless they also have the same functional groups. This should not be confused with a nuclear Isomer which involves a nucleus at different states of excitement.
2-methylpentane	2-Methylpentane is a branched-chain alkane with the molecular formula C_6H_{14}. It is a structural isomer of hexane composed of a methyl group bonded to the second carbon atom in a pentane chain. It is of similar structure to one of its isomers, 3-methylpentane, which has the methyl group located on the third carbon of the pentane chain.
Cresols	Cresols are organic compounds which are methylphenols. They are a widely occurring natural and manufactured group of aromatic organic compounds which are categorized as phenols (sometimes called phenolics.) Depending on the temperature, Cresols can be solid or liquid because they have melting points not far from room temperature.
Ethane	Ethane is a chemical compound with chemical formula C_2H_6. It is the only two-carbon alkane that is an aliphatic hydrocarbon. At standard temperature and pressure, Ethane is a colorless, odorless gas.
Ethylene	Ethylene is the chemical compound with the formula C_2H_4. It is the simplest alkene. Because it contains a carbon-carbon double bond, Ethylene is called an unsaturated hydrocarbon or an olefin.
Ethylene glycol	Ethylene glycol is an organic compound widely used as an automotive antifreeze and a precursor to polymers. In its pure form, it is an odorless, colorless, syrupy, sweet tasting liquid. Ethylene glycol was first prepared in 1859 by the French chemist Charles-Adolphe Wurtz from Ethylene glycol diacetate via saponification with potassium hydroxide and, in 1860, from the hydration of ethylene oxide.
Resorcinol	Resorcinol is a chemical compound from the dihydroxy phenols. It is the 1,3-isomer of benzenediol, and is also known with a variety of other names, including: m-dihydroxybenzene, 1,3-benzenediol, 1,3-dihydroxybenzene, 3-hydroxyphenol, m-hydroquinone, m-benzenediol, and 3-hydroxycyclohexadien-1-one. Benzene-1,3-diol is the name recommended by the International Union of Pure and Applied Chemistry (IUPAC) in its 1993 Recommendations for the Nomenclature of Organic Chemistry.
Cracking	In petroleum geology and chemistry, Cracking is the process whereby complex organic molecules such as kerogens or heavy hydrocarbons are broken down into simpler molecules (e.g. light hydrocarbons) by the breaking of carbon-carbon bonds in the precursors. The rate of Cracking and the end products are strongly dependent on the temperature and presence of any catalysts. Cracking, also referred to as pyrolysis, is the breakdown of a large alkane into smaller, more useful alkanes and an alkene.
Dimethyl ether	Dimethyl ether is the organic compound with the formula CH_3OCH_3. The simplest ether, it is a colourless gas that is a useful precursor to other organic compounds and an aerosol propellant. Dimethyl ether is also promising as a clean-burning hydrocarbon fuel.
Ether	Ether is a class of organic compounds which contain an Ether group -- an oxygen atom connected to two (substituted) alkyl or aryl groups -- of general formula R-O-R'. A typical example is the solvent and anesthetic diethyl Ether, commonly referred to simply as 'Ether' (ethoxyethane, $CH_3\text{-}CH_2\text{-}O\text{-}CH_2\text{-}CH_3$.)

Ether molecules cannot form hydrogen bonds amongst each other, resulting in a relatively low boiling point compared to that of the analogous alcohols.

Propane

Propane is a three-carbon alkane, normally a gas, but compressible to a transportable liquid. It is derived from other petroleum products during oil or natural gas processing. It is commonly used as a fuel for engines, oxy-gas torches, barbecues, portable stoves and residential central heating.

Hydrogen

Hydrogen is the chemical element with atomic number 1. It is represented by the symbol H. At standard temperature and pressure, Hydrogen is a colorless, odorless, nonmetallic, tasteless, highly flammable diatomic gas with the molecular formula H_2. With an atomic weight of 1.007 94 u, Hydrogen is the lightest element.

Denatured alcohol

Denatured alcohol is ethanol which has been rendered toxic or otherwise undrinkable, and in some cases dyed. It is used for purposes such as fuel for spirit burners and camping stoves, and as a solvent. Traditionally, the main additive was 10% methanol, which gave rise to methylated spirit.

2-chloroethanol

2-Chloroethanol is an organochlorine compound with the formula HOCH2CH2Cl. This colorless liquid has a pleasant ether -like odor. It is miscible with water.

Chlorobenzene

Chlorobenzene is an aromatic organic compound with the chemical formula C_6H_5Cl. This a colorless, flammable liquid is a common solvent and a widely used intermediate in the manufacture of other chemicals.

Chlorobenzene once was used in the manufacture of certain pesticides, most notably DDT by reaction with chloral (trichloroacetaldehyde), but this application has declined with the diminished use of DDT. At one time, Chlorobenzene was the main precursor for the manufacture of phenol:

$$C_6H_5Cl + NaOH \rightarrow C_6H_5OH + NaCl$$

As of 2005, the major use of Chlorobenzene is as an intermediate in the production of commodities such as herbicides, dyestuffs, and rubber.

Sodium

Sodium is a metallic element with a symbol Na and atomic number 11. It is a soft, silvery-white, highly reactive metal and is a member of the alkali metals within 'group 1' (formerly known as 'group IA'.) It has only one stable isotope, ^{23}Na.

Sodium ethoxide

Sodium ethoxide is an alkoxide salt with the chemical formula C_2H_5ONa.

It is commercially available as a dry yellow solid, or as a solution in ethanol. It is easily prepared in the laboratory by reacting sodium metal with ethanol:

$$2\ C_2H_5OH + 2\ Na \rightarrow 2\ C_2H_5ONa + H_2$$

Sodium ethoxide is commonly used in the Claisen condensation and malonic ester synthesis, if an ethyl ester is one of the reactants.

Alkoxide	An Alkoxide is the conjugate base of an alcohol and therefore consists of an organic group bonded to a negatively charged oxygen atom. They can be written as RO^-, where R is the organic substituent. Alkoxide s are strong bases and, when R is not bulky, good nucleophiles and good ligands.
Cyclohexane	Cyclohexane is a cycloalkane with the molecular formula C_6H_{12}. Cyclohexane is used as a nonpolar solvent for the chemical industry, and also as a raw material for the industrial production of adipic acid and caprolactam, both of which are intermediates used in the production of nylon. On an industrial scale, Cyclohexane is produced by reacting benzene with hydrogen.
Cyclohexanol	Cyclohexanol is the organic compound with the formula $(CH_2)_5CHOH$. The molecule is related to cyclohexane ring by replacement of one hydrogen atom by a hydroxyl group. This compound exists as a deliquescent colorless solid, which, when very pure, melts near room temperature. Billions of kilograms are produced annually, mainly as a precursor to nylon.
Alkyl	In chemistry, an Alkyl group is a hydrocarbon; typically an Alkyl is a part of a larger molecule. The term is usually used loosely, there is no general formula for an Alkyl group. In structural formulae, an Alkyl group is represented with an R. Usually, Alkyl groups resemble hydrocarbons, but with one less hydrogen atom.
Halide	A Halide is a binary compound, of which one part is a halogen atom and the other part is an element or radical that is less electronegative than the halogen, to make a fluoride, chloride, bromide, iodide, or astatide compound. Many salts are Halide s. All Group 1 metals form Halide s with the halogens and they are white solids.
Electrophilic addition	In organic chemistry, an Electrophilic addition reaction is an addition reaction where, in a chemical compound, a pi bond is removed by the creation of two new covalent bonds. The substrate of an Electrophilic addition reaction must have a double bond or triple bond . The driving force for this reaction is the formation of an electrophile X^+ that forms a covalent bond with an electron-rich unsaturated C=C bond.
Reagent	A Reagent or reactant is a substance or compound consumed during a chemical reaction. Solvents and catalysts, although they are involved in the reaction, are usually not referred to as reactants. Although the terms reactant and Reagent are often used interchangeably, a Reagent is more specifically 'a test substance that is added to a system in order to bring about a reaction or to see whether a reaction occurs'.
Iodide	An Iodide ion is an iodine atom with a −1 charge. Compounds with iodine in formal oxidation state −1 are called Iodide s. This can include ionic compounds such as caesium Iodide or covalent compounds such as phosphorus tri Iodide .
Allylmagnesium bromide	Allylmagnesium bromide is an Grignard reagent used for introducing the allyl group. It is commonly available as a solution in diethyl ether. If desired, it may be synthesized in the normal method from magnesium and allyl bromide.
Organolithium reagent	An Organolithium reagent is an organometallic compound with a direct bond between a carbon and a lithium atom. As the electropositive nature of lithium puts most of the charge density of the bond on the carbon atom, effectively creating a carbanion, organolithium compounds are extremely powerful bases and nucleophiles.

Organolithium reagents are industrially prepared by the reaction of an halocarbon with lithium metal, i.e. R-X + 2 Li → R-Li + LiX. A side reaction of this synthesis, especially with alkyl iodides, is the Wurtz reaction, in which an R-Li species reacts with an R-X species forming an R-R coupled product.

N-butyllithium

N-Butyllithium is the most prominent organolithium reagent. It enjoys wide use as a polymerisation initiator in the production of elastomers such as polybutadiene or styrene-butadiene-styrene (SBS.) Also, it is broadly employed as a strong base (superbase) in organic synthesis, both industrially and in the laboratory.

Formaldehyde

Formaldehyde (IUPAC name methanal) is a chemical compound with the formula CH_2O. It is the simplest aldehyde. Formaldehyde also exists as the cyclic trimer trioxane and the polymer para Formaldehyde . It exists in water as the hydrate $H_2C(OH)_2$.

Ketone

In organic chemistry, a Ketone is a type of compound that features a carbonyl group bonded to two other carbon atoms, i.e., R_3CCO-CR_3 where R can be a variety of atoms and groups of atoms. With carbonyl carbon bonded to two carbon atoms, Ketone s are distinct from many other functional groups, such as carboxylic acids, aldehydes, esters, amides, and other oxygen-containing compounds. The double-bond of the carbonyl group distinguishes Ketone s from alcohols and ethers.

1-pentanol

1-Pentanol, (or n-pentanol, pentan-1-ol), is an alcohol with five carbon atoms and the molecular formula $C_5H_{12}O$. 1-Pentanol is a colorless liquid with an unpleasant aroma. There are 7 other structural isomers of pentanol

2-butanol

2-Butanol, or sec-butanol, is a organic compound with formula $CH_3CH(OH)CH_2CH_3$. This secondary alcohol is a flammable, colorless liquid that is soluble in 12 parts water and completely miscible with polar organic solvent such as ethers and other alcohols. It is produced on a large scale, primarily as a precursor to the industrial solvent methyl ethyl ketone.

Esters

Esters are chemical compounds derived formally from an oxoacid (one containing an oxo group, X=O), and a hydroxyl compound such as an alcohol or phenol. Esters consist of an inorganic acid or organic acid in which at least one -OH (hydroxyl) group is replaced by an -O-alkyl (alkoxy) group. They are analogous to salts, using organic alcohols instead of metallic hydroxides.

Ethylmagnesium bromide

Ethylmagnesium bromide is an organomagnesium compound. Apart from acting as the synthetic equivalent of an ethyl anion synthon for nucleophilic addition, it may be used as a strong base to deprotonate various substrates such as alkynes:

RC≡CH + EtMgBr → RC≡CMgBr + EtH

This application has been supplanted by the wide availability of organolithium reagents.

Ethylmagnesium bromide is commercially available, usually as a solution in diethyl ether or tetrahydrofuran.

Epoxide

An Epoxide is a cyclic ether with three ring atoms. This ring approximately defines an equilateral triangle, which makes it highly strained. The strained ring makes Epoxide s more reactive than other ethers.

Ethylene oxide

Ethylene oxide, also called oxirane, is the organic compound with the formula C_2H_4O. This colorless flammable gas with a faintly sweet odor is the simplest epoxide, a three-membered ring consisting of two carbon and one oxygen atom. It is commonly handled and shipped as a refrigerated liquid. It is the chief precursor to ethylene glycol and other high-volume chemicals but is also used for medical sterilization.

Gilman reagent

A Gilman reagent is a lithium and copper (diorganocopper) reagent compound, R_2CuLi, where R is an organic radical. These are useful because they react with chlorides, bromides, and iodides to replace the halide group with an R group. This is extremely useful in creating larger molecules from smaller ones.

Sodium borohydride

Sodium borohydride, also known as sodium tetrahydroborate, has the chemical formula $NaBH_4$. This white solid, usually encountered as a powder, is a specialty reducing agent used in the manufacture of pharmaceuticals and other organic and inorganic compounds. It is soluble in methanol and water, but reacts with both in the absence of base.

Cyclopentanone

Cyclopentanone is a colorless liquid organic compound with a peppermint-like odor. It is a cyclic ketone, structurally similar to cyclopentane, consisting of a five-membered ring containing a ketone functional group.

The compound is stable, but flammable; the vapour is denser than air by 2.3 times.

Hydrogenation

Hydrogenation is the chemical reaction that results from the addition of hydrogen (H_2.) The process is usually employed to reduce or saturate organic compounds. The process typically constitutes the addition of pairs of hydrogen atoms to a molecule.

Nucleophilic addition

In organic chemistry, a Nucleophilic addition reaction is an addition reaction where in a chemical compound a π bond is removed by the creation of two new covalent bonds by the addition of a nucleophile .

Addition reactions are limited to chemical compounds that have multiple-bonded atoms:

- molecules with carbon - hetero multiple bonds like carbonyls, imines or nitriles
- molecules with carbon - carbon double bonds or triple bonds

Addition reactions of a nucleophile to carbon - hetero double bonds such as C=O or CN triple bond show a wide variety. These bonds are polar (have a large difference in electronegativity between the two atoms) consequently carbon carries a partial positive charge. This makes this atom the primary target for the nucleophile.

2-mercaptoethanol

2-Mercaptoethanol is the chemical compound with the formula $HOCH_2CH_2SH$. It is a hybrid of ethylene glycol, $HOCH_2CH_2OH$, and 1,2-ethanedithiol, $HSCH_2CH_2SH$. ME or βME, as it is commonly abbreviated, is used to reduce disulfide bonds and can act as a biological antioxidant by scavenging hydroxyl radicals (amongst others.) It is widely used because the hydroxyl group confers solubility in water and lowers the volatility. Due to its diminished vapor pressure, its odour, while unpleasant, is less objectionable than related thiols.

Butane

Butane also called n Butane is the unbranched alkane with four carbon atoms, $CH_3CH_2CH_2CH_3$. Butane is also used as a collective term for n Butane together with its only other isomer, iso Butane (also called methylpropane), $CH(CH_3)_3$.

Butane s are highly flammable, colorless, odorless, easily liquefied gases.

Methanethiol	Methanethiol is a colorless gas with a smell like rotten cabbage. It is a natural substance found in the blood, brain, and other animal as well as plant tissues. It is disposed of through animal feces.
Thiol	In organic chemistry, a Thiol is a compound that contains the functional group composed of a sulfur atom and a hydrogen atom (-SH.) Being the sulfur analogue of an alcohol group (-OH), this functional group is referred to either as a Thiol group or a sulfhydryl group. More traditionally, thiols are often referred to as mercaptans.
Allicin	Allicin is an organic compound obtained from garlic. It is also obtainable from onions, and other species in the family Alliaceae. It was first isolated and studied in the laboratory by Chester J. Cavallito in 1944.
Benzenesulfonic acid	Benzenesulfonic acid is an organic compound with the formula $C_6H_5SO_3H$. It is the simplest aromatic sulfonic acid. It is a colourless deliquescent sheet crystal or white wax solid that is soluble in water and ethanol, slightly soluble in benzene and insoluble in carbon disulfide and diethyl ether. It is usually stored in the form of alkali metal salts.
Glutathione	Glutathione is a tripeptide. It contains an unusual peptide linkage between the amine group of cysteine and the carboxyl group of the glutamate side chain. Glutathione, an antioxidant, helps protect cells from reactive oxygen species such as free radicals and peroxides.

Grignard reaction

The Grignard reaction is an organometallic chemical reaction in which alkyl- or aryl-magnesium halides , act as nucleophiles, attack electrophilic carbon atoms that are present within polar bonds to yield a carbon-carbon bond (compare to Wittig reaction), thus altering hybridization about the reaction center. The Grignard reaction is an important tool in the formation of carbon-carbon bonds and for the formation of carbon-phosphorus, carbon-tin, carbon-silicon, carbon-boron and other carbon-heteroatom bonds.

The addition to the nucleophile is irreversible due to the high pK_a value of the alkyl component (pK_a = ~45.)

Alcohol

In chemistry, an Alcohol is any organic compound in which a hydroxyl group (-OH) is bound to a carbon atom of an alkyl or substituted alkyl group. The general formula for a simple acyclic Alcohol is $C_nH_{2n+1}OH$. In common terms, the word Alcohol refers to ethanol, the type of Alcohol found in Alcohol ic beverages.

Ethanol is a colorless, volatile liquid with a mild odor which can be obtained by the fermentation of sugars.

Chromic acid

Chromic acid generally refers to a collection of compounds generated by the acidification of solutions containing chromate and dichromate anions or the dissolving of chromium trioxide in sulfuric acid. Often the species are assigned the formulas H_2CrO_4 and $H_2Cr_2O_7$. The anhydride of these ' Chromic acid s' is chromium trioxide, also called chromium(VI) oxide; industrially, this compound is sometimes sold as Chromic acid '

Regardless of its exact formula, Chromic acid features chromium in an oxidation state of +6 (or VI), often referred to as hexavalent chromium.

Cyclohexane

Cyclohexane is a cycloalkane with the molecular formula C_6H_{12}. Cyclohexane is used as a nonpolar solvent for the chemical industry, and also as a raw material for the industrial production of adipic acid and caprolactam, both of which are intermediates used in the production of nylon. On an industrial scale, Cyclohexane is produced by reacting benzene with hydrogen.

Reagent

A Reagent or reactant is a substance or compound consumed during a chemical reaction. Solvents and catalysts, although they are involved in the reaction, are usually not referred to as reactants.

Although the terms reactant and Reagent are often used interchangeably, a Reagent is more specifically 'a test substance that is added to a system in order to bring about a reaction or to see whether a reaction occurs'.

Alkyl

In chemistry, an Alkyl group is a hydrocarbon; typically an Alkyl is a part of a larger molecule. The term is usually used loosely, there is no general formula for an Alkyl group. In structural formulae, an Alkyl group is represented with an R. Usually, Alkyl groups resemble hydrocarbons, but with one less hydrogen atom.

Halide

A Halide is a binary compound, of which one part is a halogen atom and the other part is an element or radical that is less electronegative than the halogen, to make a fluoride, chloride, bromide, iodide, or astatide compound. Many salts are Halide s. All Group 1 metals form Halide s with the halogens and they are white solids.

Carboxylic acids

Carboxylic acids are organic acids characterized by the presence of a carboxyl group, which has the formula -C(=O)OH, usually written -COOH or $-CO_2H$. Carboxylic acids are Brønsted-Lowry acids -- they are proton donors. Salts and anions of Carboxylic acids are called carboxylates.

The simplest series of Carboxylic acids are the alkanoic acids, R-COOH, where R is a hydrogen or an alkyl group.

Esters

Esters are chemical compounds derived formally from an oxoacid (one containing an oxo group, X=O), and a hydroxyl compound such as an alcohol or phenol. Esters consist of an inorganic acid or organic acid in which at least one -OH (hydroxyl) group is replaced by an -O-alkyl (alkoxy) group. They are analogous to salts, using organic alcohols instead of metallic hydroxides.

1-heptanol

1-Heptanol is an alcohol with a seven carbon chain and the structural formula of $CH_3(CH_2)_6OH$. It is a clear colorless liquid that is very slightly soluble in water, but miscible with ether and ethanol.

There are three other isomers of heptanol that have a straight chain, 2-heptanol, 3-heptanol and 4-heptanol, which differ by the location of the alcohol functional group.

Hemiacetals

Hemiacetals and hemiketals are compounds of the general formula $R_1R'_1C(OH)OR_2$, where R_2 is not hydrogen. These types of compounds are formed from carbonyl compounds and alcohols, i.e. Hemiacetals from aldehydes, hemiketals from ketones. Thus, for Hemiacetals, either R_1 or R'_1 must be a hydrogen, while for hemiketals, neither will be.

Pyridinium

Pyridinium refers to the cationic form of pyridine. This can either be due to protonation of the ring nitrogen or because of addition of a substituent to the ring nitrogen, typically via alkylation. The lone pair of electrons on the nitrogen atom of pyridine is not delocalized, and thus pyridine can be protonated easily.

Pyridinium chlorochromate

Pyridinium chlorochromate is a reddish orange solid reagent used to oxidize primary alcohols to aldehydes and secondary alcohols to ketones. Pyridinium chlorochromate, or Pyridinium chlorochromateC, will not fully oxidize the alcohol to the carboxylic acid as does the Jones reagent. A disadvantage to using Pyridinium chlorochromateC is its toxicity.

Methanol

Methanol, also known as methyl alcohol, carbinol, wood alcohol, wood naphtha or wood spirits, is a toxic chemical with chemical formula CH_3OH Drinking even small amounts can cause blindness. It is the simplest alcohol, and is a light, volatile, colourless, flammable, toxic liquid with a distinctive odor that is very similar to but slightly sweeter than ethanol (drinking alcohol.)

Collins reagent

Collins reagent is the complex of chromium(VI) oxide with pyridine in dichloromethane. It is used to selectively oxidize primary alcohols to the aldehyde, and will tolerate many other functional groups within the molecule. It can be used as an alternative to the Jones reagent and pyridinium chlorochromate when oxidising secondary alcohols to ketones.

Dehydrogenation

Dehydrogenation is a chemical reaction that involves the elimination of hydrogen (H_2.) It is the reverse process of hydrogenation. Dehydrogenation reactions may be either large scale industrial processes or smaller scale laboratory procedures.

Jones reagent

The Jones oxidation is a chemical reaction described as the chromic acid oxidation of primary and secondary alcohols to carboxylic acids and ketones, respectively. Jones reagent - a solution of chromium trioxide in concentrated sulfuric acid - is used as the oxidizing agent.

The Jones reagent will also completely oxidize aldehydes to carboxylic acids.

Potassium permanganate	Potassium permanganate is the inorganic chemical compound $KMnO_4$, a water soluble salt consisting of equal mole amounts of potassium (K^+) and permanganate (MnO_4^-, officially called manganate (VII)) ions. This salt, formerly known as permanganate of potash or Condy's crystals is a strong oxidizing agent. It dissolves in water to give deep purple solutions, evaporation of which gives prismatic purplish-black glistening crystals.
Swern oxidation	The Swern oxidation is a chemical reaction whereby a primary or secondary alcohol is oxidized to an aldehyde or ketone using oxalyl chloride, dimethyl sulfoxide (DMSwern oxidation) and an organic base, such as triethylamine. The reaction is known for its mild character and wide tolerance of functional groups. The by-products are dimethyl sulfide (Me_2S), carbon monoxide (CO), carbon dioxide (CO_2) and -- when triethylamine is used as base -- triethylammonium chloride (Et_3NHCl.)
Nicotinamide	Nicotinamide, also known as niacinamide and nicotinic acid amide, is the amide of nicotinic acid (vitamin B_3.) Nicotinamide is a water-soluble vitamin and is part of the vitamin B group. Nicotinic acid, also known as niacin, is converted to niacinamide in vivo, and though the two are identical in their vitamin functions, niacinamide does not have the same pharmacologic and toxic effects of niacin, which occur incidental to niacin's conversion.
Nicotinamide adenine dinucleotide	Nicotinamide adenine dinucleotide, abbreviated Nicotinamide adenine dinucleotide$^+$, is a coenzyme found in all living cells. The compound is a dinucleotide, since it consists of two nucleotides joined through their phosphate groups: with one nucleotide containing an adenine base, and the other containing nicotinamide. In metabolism, Nicotinamide adenine dinucleotide$^+$ is involved in redox reactions, carrying electrons from one reaction to another.
Aldehyde	An Aldehyde is an organic compound containing a terminal carbonyl group. This functional group, which consists of a carbon atom bonded to a hydrogen atom and double-bonded to an oxygen atom (chemical formula O=CH-), is called the Aldehyde group. The Aldehyde group is also called the formyl or methanoyl group.
Ethylene	Ethylene is the chemical compound with the formula C_2H_4. It is the simplest alkene. Because it contains a carbon-carbon double bond, Ethylene is called an unsaturated hydrocarbon or an olefin.
Ethylene glycol	Ethylene glycol is an organic compound widely used as an automotive antifreeze and a precursor to polymers. In its pure form, it is an odorless, colorless, syrupy, sweet tasting liquid. Ethylene glycol was first prepared in 1859 by the French chemist Charles-Adolphe Wurtz from Ethylene glycol diacetate via saponification with potassium hydroxide and, in 1860, from the hydration of ethylene oxide.
Formaldehyde	Formaldehyde (IUPAC name methanal) is a chemical compound with the formula CH_2O. It is the simplest aldehyde. Formaldehyde also exists as the cyclic trimer trioxane and the polymer para Formaldehyde . It exists in water as the hydrate $H_2C(OH)_2$.
Formic acid	Formic acid is the simplest carboxylic acid. Its formula is HCOOH or CH_2O_2. It is an important intermediate in chemical synthesis and occurs naturally, most notably in the venom of bee and ant stings.

Niacin

Niacin, also known as vitamin B_3 or nicotinic acid, is a water-soluble vitamin that prevents the deficiency disease pellagra. It is an organic compound with the molecular formula $C_6H_5NO_2$. It is a derivative of pyridine, with a carboxyl group (COOH) at the 3-position.

Nicotinic acid

Nicotinic acids are derivatives of pyridine which have a substituted carboxy group.

Although the term niacin is sometimes used interchangeably with 'Nicotinic acid', it is more precise to only apply it to the 1-3 (meta) substituted form.

Examples include:

- arecoline
- nicotinamide
- nicorandil
- nikethamide
- nimodipine

.

Oxalic acid

Oxalic acid is the chemical compound with the formula $H_2C_2O_4$. This dicarboxylic acid is better described with the formula HOOCCOOH. It is a relatively strong organic acid, being about 3,000 times as strong as acetic acid. The di-anion, known as oxalate, is also a reducing agent as well as a ligand in coordination chemistry.

Electrophile

In chemistry, an Electrophile (literally electron-lover) is a reagent attracted to electrons that participates in a chemical reaction by accepting an electron pair in order to bond to a nucleophile. Because Electrophile s accept electrons, they are Lewis acids Most Electrophile s are positively charged, have an atom which carries a partial positive charge, or have an atom which does not have an octet of electrons.

Nucleophile

In chemistry, a Nucleophile is a reagent that forms a chemical bond to its reaction partner (the electrophile) by donating both bonding electrons. Because nucleophiles donate electrons, they are by definition Lewis bases All molecules or ions with a free pair of electrons can act as nucleophiles.

Butyl

In organic chemistry, Butyl is a four-carbon alkyl substituent with chemical formula $-C_4H_9$. It is derived from either of the two isomers of the alkane called butane.

Each of the two isomers of butane give rise to two isomers of the Butyl substituent.

Cycloalkanes

Cycloalkanes are types of alkanes which have one or more rings of carbon atoms in the chemical structure of their molecules. Alkanes are types of organic hydrocarbon compounds which have only single chemical bonds in their chemical structure. Cycloalkanes consist of only carbon and hydrogen atoms and are saturated because there are no multiple C-C bonds to hydrogenate

Cyclopentanol

Cyclopentanol or cyclopentyl alcohol is a cyclic alcohol.

The dehydration of Cyclopentanol produces cyclopentene:

$C_5H_{10}O \rightarrow C_5H_8 + H_2O$

.

Protonation

In chemistry, Protonation is the addition of a proton (H^+) to an atom, molecule, or ion. Protonation is possibly the most fundamental chemical reaction and is a step in many stoichiometric and catalytic processes. Some ions and molecules can undergo more than one Protonation and are labeled polybasic.

Bromide

A Bromide ion is a bromine atom with charge of −1.

Compounds with bromine in formal oxidation state −1 are called Bromide s, and each individual chemical in this class can be called a Bromide as well. The class name can include ionic compounds such as caesium Bromide or covalent compounds such as sulfur di Bromide .

Butylated hydroxyanisole

Butylated hydroxyanisole is an antioxidant consisting of a mixture of two isomeric organic compounds, 2-tert-butyl-4-hydroxyanisole and 3-tert-butyl-4-hydroxyanisole. It is prepared from 4-methoxyphenol and isobutylene. It is a waxy solid used in certain amounts as a food additive with the E number E320.

Sodium

Sodium is a metallic element with a symbol Na and atomic number 11. It is a soft, silvery-white, highly reactive metal and is a member of the alkali metals within 'group 1' (formerly known as 'group IA'.) It has only one stable isotope, ^{23}Na.

Zinc

Zinc is a metallic chemical element with the symbol Zn and atomic number 30. It is a first-row transition metal in group 12 of the periodic table. Zinc is chemically similar to magnesium because its ion is of similar size and its only common oxidation state is +2.

Neopentyl alcohol

Neopentyl alcohol is a compound with formula $(CH_3)_3CCH_2OH$. .

Phosphorus halides

There are three series of binary Phosphorus halides, containing phosphorus in the oxidation states +5, +3 and +2. All twelve compounds have been described, in varying degrees of detail, although serious doubts have been cast on the existence of PI_5.

Phosphorus pentafluoride is a relatively inert gas, notable as a mild Lewis acid and a fluoride ion acceptor.

Dehydration reaction

In chemistry, a Dehydration reaction is usually defined as a chemical reaction that involves the loss of water from the reacting molecule. Dehydration reaction s are a subset of elimination reactions. Because the hydroxyl group (-OH) is a poor leaving group, having an Brønsted acid catalyst often helps by protonating the hydroxyl group to give the better leaving group, $-OH_2^+$.

Phosphorus tribromide

Phosphorus tribromide is a colourless liquid with the formula PBr_3. It fumes in air due to hydrolysis and has a penetrating odour. It is widely used in the laboratory for the conversion of alcohols to alkyl bromides.

Thionyl chloride

Thionyl chloride is an inorganic compound with the formula $SOCl_2$. It is a reactive chemical reagent used in chlorination reactions. It is a colorless, distillable liquid at room temperature and pressure that decomposes above 140 °C. Thionyl chloride is sometimes confused with sulfuryl chloride, SO_2Cl_2, but the chemical properties of these S(IV) and S(VI) compounds differ significantly.

Octane

Octane is a straight-chain alkane with the chemical formula $CH_3(CH_2)_6CH_3$.

Octane has 18 structural isomers (25 including stereoisomers):

- Octane
- 2-Methylheptane
- 3-Methylheptane (2 enantiomers)
- 4-Methylheptane
- 3-Ethylhexane
- 2,2-Dimethylhexane
- 2,3-Dimethylhexane (2 enantiomers)
- 2,4-Dimethylhexane (2 enantiomers)
- 2,5-Dimethylhexane
- 3,3-Dimethylhexane
- 3,4-Dimethylhexane (2 enantiomers + 1 meso compound)
- 3-Ethyl-2-methylpentane (2 enantiomers)
- 3-Ethyl-3-methylpentane
- 2,2,3-Trimethylpentane (2 enantiomers)
- 2,2,4-Trimethylpentane (isooctane)
- 2,3,3-Trimethylpentane
- 2,3,4-Trimethylpentane
- 2,2,3,3-Tetramethylbutane

Octane became well-known in American popular culture in the mid- and late-sixties, when gasoline companies boasted of 'high Octane' levels in their gasoline on television commercials. These commercials are referring to the Octane rating, which is a measure for the anti-knocking properties of gasoline. The Octane rating is not related to the amount of Octane contained in the gasoline.

Alkene

In organic chemistry, an Alkene olefin, or olefine is an unsaturated chemical compound containing at least one carbon-to-carbon double bond. The simplest acyclic Alkene s, with only one double bond and no other functional groups, form a homologous series of hydrocarbons with the general formula C_nH_{2n}.

The simplest Alkene is ethylene (C_2H_4), which has the International Union of Pure and Applied Chemistry (IUPAC) name ethene.

Methylcyclohexane

Methylcyclohexane is a colourless liquid with a faint benzene-like odour. Its molecular formula is C_7H_{14}. Methylcyclohexane is used in organic synthesis and as a solvent for cellulose ethers.

Ether	Ether is a class of organic compounds which contain an Ether group -- an oxygen atom connected to two (substituted) alkyl or aryl groups -- of general formula R-O-R'. A typical example is the solvent and anesthetic diethyl Ether, commonly referred to simply as 'Ether' (ethoxyethane, CH_3-CH_2-O-CH_2-CH_3.) Ether molecules cannot form hydrogen bonds amongst each other, resulting in a relatively low boiling point compared to that of the analogous alcohols.
Diol	A Diol or glycol is a chemical compound containing two hydroxyl groups (-OH groups) Vicinal Diol s have hydroxyl groups attached to adjacent atoms. Examples of vicinal Diol compounds are ethylene glycol and propylene glycol. Geminal Diol s have hydroxyl groups bonded to the same atom.
Pinacol	Pinacol is a white solid organic compound. It may be produced by the Pinacol coupling reaction from acetone. As a vicinal-diol, it can rearrange to pinacolone by the Pinacol rearrangement.
Pinacol rearrangement	The Pinacol rearrangement or pinacol-pinacolone rearrangement is a method for converting a 1,2-diol to a carbonyl compound in organic chemistry. This rearrangement takes place under acidic conditions. The name of the reaction comes from the rearrangement of pinacol to pinacolone.
Esterification	Esterification is the general name for a chemical reaction in which two reactants (typically an alcohol and an acid) form an ester as the reaction product. Esters are common in organic chemistry and biological materials, and often have a characteristic pleasant, fruity odor. This leads to their extensive use in the fragrance and flavor industry.
Claisen condensation	The Claisen condensation is a carbon-carbon bond forming reaction that occurs between two esters or one ester and another carbonyl compound in the presence of a strong base, resulting in a β-keto ester or a β-diketone. It is named after Rainer Ludwig Claisen, who first published his work on the reaction in 1881 . At least one of the reagents must be enolizable (have an α-proton and be able to undergo deprotonation to form the enolate anion.)
Methylamine	Methylamine is the organic compound with a formula of CH_3NH_2. This colourless gas is a derivative of ammonia, wherein one H atom is replaced by a methyl group. It is the simplest primary amine.
Nitrate	In inorganic chemistry, a Nitrate is a salt of nitric acid with an ion composed of one nitrogen and three oxygen atoms (NO_3^- .) In organic chemistry the esters of nitric acid and various alcohols are called nitrates. The Nitrate ion is a polyatomic ion with the empirical formula NO_3^- and a molecular mass of 62.0049.
Nitroglycerin	Nitroglycerin, trinitroglycerin, trinitroglycerine, 1,2,3-trinitroxypropane and glyceryl trinitrate, is a heavy, colorless, oily, explosive liquid obtained by nitrating glycerol. Since the 1860s, it has been used as an active ingredient in the manufacture of explosives, specifically dynamite, and as such is employed in the construction and demolition industries. Similarly, since the 1880s, it has been used by the military as an active ingredient, and a gellatinizer for nitrocellulose, in some solid propellants, such as Cordite and Ballistite.

Phosphate	A Phosphate, an inorganic chemical, is a salt of phosphoric acid. Inorganic phosphates are mined to obtain phosphorus for use in agriculture and industry. In organic chemistry, a Phosphate, or organophosphate, is an ester of phosphoric acid.
Picric acid	Picric acid is the chemical compound more formally called 2,4,6-trinitrophenol (TNP.) This, a yellow crystalline solid, is one of the most acidic phenols. Like other highly nitrated compounds such as TNT, Picric acid is an explosive.
Trimethyl phosphate	Trimethyl phosphate, $(CH_3O)_3PO$, is the trimethyl ester of phosphoric acid. Trimethyl phosphate is a methylating agent for nitrogen heterocyclic compounds. It is used as a color inhibitor for fibers (e.g. polyester) and other polymers.
Trinitrotoluene	Trinitrotoluene , or more specifically, 2,4,6-Trinitrotoluene, is a chemical compound with the formula $C_6H_{23}CH_3$. This yellow -coloured solid is a reagent (reactant) in chemistry but is best known as a useful explosive material with convenient handling properties. The explosive yield of TNT is considered the standard measure of strength of bombs and other explosives.
Pentaerythritol	Pentaerythritol is the organic compound with the formula $C(CH_2OH)_4$. This white, crystalline polyol is a versatile building block for the preparation of many polyfunctionalized compounds such as the explosive PETN and Pentaerythritol triacrylate. Derivatives of Pentaerythritol are components of alkyd resins, varnishes, PVC stabilizers, tall oil esters, and olefin antioxidants.
Alkoxide	An Alkoxide is the conjugate base of an alcohol and therefore consists of an organic group bonded to a negatively charged oxygen atom. They can be written as RO^-, where R is the organic substituent. Alkoxide s are strong bases and, when R is not bulky, good nucleophiles and good ligands.
Williamson ether synthesis	The Williamson ether synthesis is an organic reaction, forming an ether from a organohalide and an alcohol. This reaction was developed by Alexander Williamson in 1850 . Typically it involves the reaction of an alkoxide ion with a primary alkyl halide via an S_N2 reaction.

Methanol	Methanol, also known as methyl alcohol, carbinol, wood alcohol, wood naphtha or wood spirits, is a toxic chemical with chemical formula CH_3OH Drinking even small amounts can cause blindness. It is the simplest alcohol, and is a light, volatile, colourless, flammable, toxic liquid with a distinctive odor that is very similar to but slightly sweeter than ethanol (drinking alcohol.)
Hydrogen	Hydrogen is the chemical element with atomic number 1. It is represented by the symbol H. At standard temperature and pressure, Hydrogen is a colorless, odorless, nonmetallic, tasteless, highly flammable diatomic gas with the molecular formula H_2. With an atomic weight of 1.007 94 u, Hydrogen is the lightest element.
Octane	Octane is a straight-chain alkane with the chemical formula $CH_3(CH_2)_6CH_3$. Octane has 18 structural isomers (25 including stereoisomers): • Octane • 2-Methylheptane • 3-Methylheptane (2 enantiomers) • 4-Methylheptane • 3-Ethylhexane • 2,2-Dimethylhexane • 2,3-Dimethylhexane (2 enantiomers) • 2,4-Dimethylhexane (2 enantiomers) • 2,5-Dimethylhexane • 3,3-Dimethylhexane • 3,4-Dimethylhexane (2 enantiomers + 1 meso compound) • 3-Ethyl-2-methylpentane (2 enantiomers) • 3-Ethyl-3-methylpentane • 2,2,3-Trimethylpentane (2 enantiomers) • 2,2,4-Trimethylpentane (isooctane) • 2,3,3-Trimethylpentane • 2,3,4-Trimethylpentane • 2,2,3,3-Tetramethylbutane Octane became well-known in American popular culture in the mid- and late-sixties, when gasoline companies boasted of 'high Octane' levels in their gasoline on television commercials. These commercials are referring to the Octane rating, which is a measure for the anti-knocking properties of gasoline. The Octane rating is not related to the amount of Octane contained in the gasoline.
Carbon-carbon bond	A Carbon-carbon bond is a covalent bond between two carbon atoms. The most common form is the single bond - a bond composed of two electrons, one from each of the two atoms. The carbon-carbon single bond is a sigma bond and is said to be formed between one hybridized orbital from each of the carbon atoms.
Cyclohexene	Cyclohexene is a colorless clear liquid cycloalkene with an intense aversive characteristic sharp smell reminiscent of an oil refinery.

It is not very stable upon long term storage with exposure to light and air and should be distilled before use to eliminate peroxides. A common experiment among beginning organic chemistry students is the acid catalyzed dehydration of cyclohexanol with distillative removal of the resulting Cyclohexene from the reaction mixture:

- Refractive Index: 1.4465
- Critical temperature: 287.2 °C (560.4 K)

Hydrocarbon

In organic chemistry, a Hydrocarbon is an organic compound consisting entirely of hydrogen and carbon. With relation to chemical terminology, aromatic Hydrocarbon s or arenes, alkanes, alkenes and alkyne-based compounds composed entirely of carbon or hydrogen are referred to as 'pure' Hydrocarbon s, whereas other Hydrocarbon s with bonded compounds or impurities of sulfur or nitrogen, are referred to as 'impure', and remain somewhat erroneously referred to as Hydrocarbon s.

Hydrocarbon s are referred to as consisting of a 'backbone' or 'skeleton' composed entirely of carbon and hydrogen and other bonded compounds, and have a functional group that generally facilitates combustion.

Carbon-hydrogen bonds

A carbogen is a single bond between carbon and hydrogen atoms, most commonly found in organic compounds . Carbon-hydrogen bonds have a bond length of 1.09 Å and a bond energy of 413 kJ/mol. Using Pauling's scale--C and H--the electronegativity difference between these two atoms is 0.4.

Newman projection

A Newman projection, useful in alkane stereochemistry, visualizes chemical conformations of a carbon-carbon chemical bond from front to back, with the front carbon represented by a dot and the back carbon as a circle The front carbon atom is called proximal, while the back atom is called distal. This type of representation is useful for assessing the torsional angle between bonds.

Alcohol

In chemistry, an Alcohol is any organic compound in which a hydroxyl group (-OH) is bound to a carbon atom of an alkyl or substituted alkyl group. The general formula for a simple acyclic Alcohol is $C_nH_{2n+1}OH$. In common terms, the word Alcohol refers to ethanol, the type of Alcohol found in Alcohol ic beverages.

Ethanol is a colorless, volatile liquid with a mild odor which can be obtained by the fermentation of sugars.

Amine

Amine s are organic compounds and functional groups that contain a basic nitrogen atom with a lone pair. Amine s are derivatives of ammonia, wherein one or more hydrogen atoms have been replaced by a substituent such as an alkyl or aryl group. Important Amine s include amino acids, biogenic Amine s, trimethyl Amine , and aniline; see Category: Amine s for a list of Amine s.

Benzoic acid

Benzoic acid, $C_7H_6O_2$ (or C_6H_5COOH), is a colorless crystalline solid and the simplest aromatic carboxylic acid. The name derived from gum benzoin, which was for a long time the only source for Benzoic acid. This weak acid and its salts are used as a food preservative.

Butyramide

Butyramide is the amide of butyric acid. It has the molecular formula $C_3H_7CONH_2$. It is a clear liquid that is highly soluble in water and ethanol, but slightly soluble in diethyl ether.

Fischer projection	The Fischer projection, devised by Hermann Emil Fischer in 1891, is a two-dimensional representation of a three-dimensional organic molecule by projection. They are used by chemists, particularly in organic chemistry and biochemistry. All bonds are depicted as horizontal or vertical lines.
Isomer	In chemistry, Isomer s are compounds with the same molecular formula but different structural formula. Isomer s do not necessarily share similar properties unless they also have the same functional groups. This should not be confused with a nuclear Isomer which involves a nucleus at different states of excitement.
Esters	Esters are chemical compounds derived formally from an oxoacid (one containing an oxo group, X=O), and a hydroxyl compound such as an alcohol or phenol. Esters consist of an inorganic acid or organic acid in which at least one -OH (hydroxyl) group is replaced by an -O-alkyl (alkoxy) group. They are analogous to salts, using organic alcohols instead of metallic hydroxides.
Ethylene	Ethylene is the chemical compound with the formula C_2H_4. It is the simplest alkene. Because it contains a carbon-carbon double bond, Ethylene is called an unsaturated hydrocarbon or an olefin.
Radicals	In chemistry, Radicals are atoms, molecules, or ions with unpaired electrons on an otherwise open shell configuration. These unpaired electrons are usually highly reactive, so Radicals are likely to take part in chemical reactions. Radicals play an important role in combustion, atmospheric chemistry, polymerization, plasma chemistry, biochemistry, and many other chemical processes, including human physiology.
Source	Source Holdings Ltd. is a specialist European-based provider of Exchange Traded Products (ETPs), including Exchange Traded Funds (ETFs) and Exchange Traded Commodities (ETCs.) The company was founded in 2008 by Bank of America Merrill Lynch, Goldman Sachs and Morgan Stanley and is headquartered inside the City of London, on 88 Wood Street. Source launched its first product offering on the 20th April 2009, although the 3 founders are reported to have started working on the project in early 2008 .
Chlorin	In organic chemistry, a Chlorin is a large heterocyclic aromatic ring consisting, at the core, of three pyrroles and one pyrroline coupled through four methine linkages. Unlike a porphyrin, a Chlorin is therefore largely aromatic but not aromatic through the entire circumference of the ring. Magnesium-containing Chlorin s are called chlorophylls, and are the central photosensitive pigment in chloroplasts.
Alkynes	Alkynes are hydrocarbons that have a triple bond between two carbon atoms, with the formula C_nH_{2n-2}. Alkynes are traditionally known as acetylenes, although the name acetylene also refers specifically to C_2H_2, known formally (but rarely) as ethyne using IUPAC nomenclature. Like other hydrocarbons, Alkynes are generally hydrophobic but tend to be more reactive.
Cycloalkanes	Cycloalkanes are types of alkanes which have one or more rings of carbon atoms in the chemical structure of their molecules. Alkanes are types of organic hydrocarbon compounds which have only single chemical bonds in their chemical structure. Cycloalkanes consist of only carbon and hydrogen atoms and are saturated because there are no multiple C-C bonds to hydrogenate

Alkene

In organic chemistry, an Alkene olefin, or olefine is an unsaturated chemical compound containing at least one carbon-to-carbon double bond. The simplest acyclic Alkene s, with only one double bond and no other functional groups, form a homologous series of hydrocarbons with the general formula C_nH_{2n}.

The simplest Alkene is ethylene (C_2H_4), which has the International Union of Pure and Applied Chemistry (IUPAC) name ethene.

Ether

Ether is a class of organic compounds which contain an Ether group -- an oxygen atom connected to two (substituted) alkyl or aryl groups -- of general formula R-O-R'. A typical example is the solvent and anesthetic diethyl Ether, commonly referred to simply as 'Ether' (ethoxyethane, CH_3-CH_2-O-CH_2-CH_3.)

Ether molecules cannot form hydrogen bonds amongst each other, resulting in a relatively low boiling point compared to that of the analogous alcohols.

Ketone

In organic chemistry, a Ketone is a type of compound that features a carbonyl group bonded to two other carbon atoms, i.e., R_3CCO-CR_3 where R can be a variety of atoms and groups of atoms. With carbonyl carbon bonded to two carbon atoms, Ketone s are distinct from many other functional groups, such as carboxylic acids, aldehydes, esters, amides, and other oxygen-containing compounds. The double-bond of the carbonyl group distinguishes Ketone s from alcohols and ethers.

Grignard reaction

The Grignard reaction is an organometallic chemical reaction in which alkyl- or aryl-magnesium halides , act as nucleophiles, attack electrophilic carbon atoms that are present within polar bonds to yield a carbon-carbon bond (compare to Wittig reaction), thus altering hybridization about the reaction center. The Grignard reaction is an important tool in the formation of carbon-carbon bonds and for the formation of carbon-phosphorus, carbon-tin, carbon-silicon, carbon-boron and other carbon-heteroatom bonds.

The addition to the nucleophile is irreversible due to the high pK_a value of the alkyl component (pK_a = ~45.)

Toluene

Toluene phenylmethane, and Toluol, is a clear water-insoluble liquid with the typical smell of paint thinners, redolent of the sweet smell of the related compound benzene. It is an aromatic hydrocarbon that is widely used as an industrial feedstock and as a solvent. Like other solvents, Toluene is also used as an inhalant drug for its intoxicating properties; however this causes severe neurological harm.

Acetylenic

In organic chemistry, the term Acetylenic designates

- a doubly unsaturated position (sp-hybridized) on a molecular framework, for instance in an alkyne such as acetylene;
- an ethynyl fragment, HC≡C-, or substituted homologue.

.

Aldehyde

An Aldehyde is an organic compound containing a terminal carbonyl group. This functional group, which consists of a carbon atom bonded to a hydrogen atom and double-bonded to an oxygen atom (chemical formula O=CH-), is called the Aldehyde group. The Aldehyde group is also called the formyl or methanoyl group.

Hydrogen

Hydrogen is the chemical element with atomic number 1. It is represented by the symbol H. At standard temperature and pressure, Hydrogen is a colorless, odorless, nonmetallic, tasteless, highly flammable diatomic gas with the molecular formula H_2. With an atomic weight of 1.007 94 u, Hydrogen is the lightest element.

Acetic acid

Acetic acid, CH_3COOH, also known as ethanoic acid, is an organic acid which gives vinegar its sour taste and pungent smell. Pure, water-free Acetic acid is a colourless liquid that absorbs water from the environment (hygroscopy), and freezes at 16.7 °C (62 °F) to a colourless crystalline solid. It is a weak acid, in that it is only partially dissociated acid in aqueous solution.

Carboxylic acids

Carboxylic acids are organic acids characterized by the presence of a carboxyl group, which has the formula -C(=O)OH, usually written -COOH or $-CO_2H$. Carboxylic acids are Brønsted-Lowry acids -- they are proton donors. Salts and anions of Carboxylic acids are called carboxylates.

The simplest series of Carboxylic acids are the alkanoic acids, R-COOH, where R is a hydrogen or an alkyl group.

Ether

Ether is a class of organic compounds which contain an Ether group -- an oxygen atom connected to two (substituted) alkyl or aryl groups -- of general formula R-O-R'. A typical example is the solvent and anesthetic diethyl Ether, commonly referred to simply as 'Ether' (ethoxyethane, $CH_3\text{-}CH_2\text{-}O\text{-}CH_2\text{-}CH_3$.)

Ether molecules cannot form hydrogen bonds amongst each other, resulting in a relatively low boiling point compared to that of the analogous alcohols.

Ethylbenzene

Ethylbenzene is an organic compound with the formula $C_6H_5CH_2CH_3$. This aromatic hydrocarbon is important in the petrochemical industry as an intermediate in the production of styrene, which in turn is used for making polystyrene, a commonly used plastic material. Although often present in small amounts in crude oil, Ethylbenzene is produced in bulk quantities by combining benzene and ethylene in an acid-catalyzed chemical reaction:

$$C_6H_6 + C_2H_4 \rightarrow C_6H_5CH_2CH_3$$

Approximately 24,700,000 tons were produced in 1999.

Norepinephrine

Noradrenaline (BAN) or Norepinephrine is a catecholamine with dual roles as a hormone and a neurotransmitter.

As a stress hormone, Norepinephrine affects parts of the brain where attention and responding actions are controlled. Along with epinephrine, Norepinephrine also underlies the fight-or-flight response, directly increasing heart rate, triggering the release of glucose from energy stores, and increasing blood flow to skeletal muscle.

Ethyl group

In chemistry, an Ethyl group is an alkyl functional group derived from ethane (C_2H_6.) It has the formula -C_2H_5 and is very often abbreviated -Et.

Ethylation is the formation of a compound by introduction of the ethyl functional group, C_2H_5.

Isopropyl

In organic chemistry, Isopropyl is a substituent form of propane, the three-carbon alkyl functional group. As an isomer of propyl, bonding to an R group occurs at the secondary (2°) carbon.

Isopropyl is also known as 1-methylethyl (IUPAC); shorthand notations for this alkyl are i-Pr,^{i}Pr, or Pr^{i} (some care has to be taken when using the abbreviation 'Pr', as it is the same as for the chemical element Praseodymium.)

Esters

Esters are chemical compounds derived formally from an oxoacid (one containing an oxo group, X=O), and a hydroxyl compound such as an alcohol or phenol. Esters consist of an inorganic acid or organic acid in which at least one -OH (hydroxyl) group is replaced by an -O-alkyl (alkoxy) group. They are analogous to salts, using organic alcohols instead of metallic hydroxides.

Alkyl

In chemistry, an Alkyl group is a hydrocarbon; typically an Alkyl is a part of a larger molecule. The term is usually used loosely, there is no general formula for an Alkyl group. In structural formulae, an Alkyl group is represented with an R. Usually, Alkyl groups resemble hydrocarbons, but with one less hydrogen atom.

Styrene

Styrene, also known as vinyl benzene as well as many other names , is an organic compound with the chemical formula $C_6H_5CH{=}CH_2$. This aromatic hydrocarbon is a colorless oily liquid that evaporates easily and has a sweet smell, although high concentrations confer a less pleasant odor. Styrene is the precursor to polystyrene and several copolymers.

Isomer

In chemistry, Isomer s are compounds with the same molecular formula but different structural formula. Isomer s do not necessarily share similar properties unless they also have the same functional groups. This should not be confused with a nuclear Isomer which involves a nucleus at different states of excitement.

Iodide

An Iodide ion is an iodine atom with a −1 charge. Compounds with iodine in formal oxidation state −1 are called Iodide s. This can include ionic compounds such as caesium Iodide or covalent compounds such as phosphorus tri Iodide .

N-propyl iodide

N-Propyl iodide is a colorless, flammable chemical compound. It has the chemical formula C_3H_7I and is prepared by heating n-propyl alcohol with iodine and phosphorus.

Allyl

An Allyl group is an alkene hydrocarbon group with the formula $H_2C{=}CH{-}CH_2{-}$. It is made up of a vinyl group, $CH_2{=}CH{-}$, attached to a methylene $-CH_2$. For example Allyl alcohol has the structure $H_2C{=}CH{-}CH_2OH$. Another example of a simple Allyl compound is Allyl chloride.

Allyl bromide

Allyl bromide is an organic halide. Its refractive index is 1.4697 (20 °C, 589 nm.) Allyl bromide is an alkylating agent used in synthesis of polymers, pharmaceuticals, allyls and other organic compounds.

Ethanol

Ethanol pure alcohol, grain alcohol is a volatile, flammable, colorless liquid. It is a psychoactive drug, best known as the type of alcohol found in alcoholic beverages and in modern thermometers. Ethanol is one of the oldest recreational drugs.

Cyclohexane

Cyclohexane is a cycloalkane with the molecular formula C_6H_{12}. Cyclohexane is used as a nonpolar solvent for the chemical industry, and also as a raw material for the industrial production of adipic acid and caprolactam, both of which are intermediates used in the production of nylon. On an industrial scale, Cyclohexane is produced by reacting benzene with hydrogen.

Hydroxyl

Hydroxyl in chemistry stands for a molecule consisting of an oxygen atom and a hydrogen atom connected by a covalent bond (single bond.) The neutral form is a Hydroxyl radical and the Hydroxyl anion is called a hydroxide. When the oxygen atom is linked to a larger molecule the Hydroxyl group is a functional group (OH) .

Fentanyl

Fentanyl -- brand names include Actiq, Duragesic, Fentora, and Sublimaze -- is a synthetic primary μ-opioid agonist commonly used to treat post-operative and chronic breakthrough pain. It is approximately 100 times more potent than morphine, with 100 micrograms of Fentanyl equivalent to 10 mg of morphine and 75 mg of pethidine in analgesic activity. It has an LD_{50} of 3.1 milligrams per kilogram in rats, 0.03 milligrams per kilogram in monkeys, and an undetermined milligrams per kilogram in humans.

Carbon

Carbon is the chemical element with symbol C and atomic number 6. As a member of group 14 on the periodic table, it is nonmetallic and tetravalent--making four electrons available to form covalent chemical bonds. There are three naturally occurring isotopes, with ^{12}C and ^{13}C being stable, while ^{14}C is radioactive, decaying with a half-life of about 5730 years.

Alpha carbon

The alpha carbon in organic chemistry refers to the first carbon that attaches to a functional group (the carbon is attached at the first position.) By extension, the second carbon is the beta carbon, and so on. This nomenclature can also be applied to the hydrogen atoms attached to the carbons.

1,1,1-trichloroethane

The chemical compound 1,1,1-Trichloroethane is a chlorinated hydrocarbon that was until recently widely used as an industrial solvent. Other names for it include methyl chloroform, chlorothene, and the trade names Solvent 111 and Genklene (used by ICI.)

1,1,1-Trichloroethane was first produced by the French chemist Henri Victor Regnault in 1840.

Diethyl ether	Diethyl ether is a clear, colorless, and highly flammable liquid with a low boiling point and a characteristic odor. It is the most common member of a class of chemical compounds known generically as ethers. It is an isomer of butanol.
Ether	Ether is a class of organic compounds which contain an Ether group -- an oxygen atom connected to two (substituted) alkyl or aryl groups -- of general formula R-O-R'. A typical example is the solvent and anesthetic diethyl Ether, commonly referred to simply as 'Ether' (ethoxyethane, CH_3-CH_2-O-CH_2-CH_3.) Ether molecules cannot form hydrogen bonds amongst each other, resulting in a relatively low boiling point compared to that of the analogous alcohols.
Tetrahydrofuran	Tetrahydrofuran is a colorless, water-miscible organic liquid with low-viscosity at standard temperature and pressure. It is a heterocyclic compound with a chemical formula C_4H_8O, and is the fully hydrogenated analog of the aromatic organic compound furan. It is one of the most polar of the organic functional class of ethers, and has a relatively low freeze point, and so is a commonly used modern organic chemical laboratory solvent across a range of temperatures.
Alcohol	In chemistry, an Alcohol is any organic compound in which a hydroxyl group (-OH) is bound to a carbon atom of an alkyl or substituted alkyl group. The general formula for a simple acyclic Alcohol is $C_nH_{2n+1}OH$. In common terms, the word Alcohol refers to ethanol, the type of Alcohol found in Alcohol ic beverages. Ethanol is a colorless, volatile liquid with a mild odor which can be obtained by the fermentation of sugars.
Dimethyl ether	Dimethyl ether is the organic compound with the formula CH_3OCH_3. The simplest ether, it is a colourless gas that is a useful precursor to other organic compounds and an aerosol propellant. Dimethyl ether is also promising as a clean-burning hydrocarbon fuel.
Hydrogen	Hydrogen is the chemical element with atomic number 1. It is represented by the symbol H. At standard temperature and pressure, Hydrogen is a colorless, odorless, nonmetallic, tasteless, highly flammable diatomic gas with the molecular formula H_2. With an atomic weight of 1.007 94 u, Hydrogen is the lightest element.
Alkene	In organic chemistry, an Alkene olefin, or olefine is an unsaturated chemical compound containing at least one carbon-to-carbon double bond. The simplest acyclic Alkene s, with only one double bond and no other functional groups, form a homologous series of hydrocarbons with the general formula C_nH_{2n}. The simplest Alkene is ethylene (C_2H_4), which has the International Union of Pure and Applied Chemistry (IUPAC) name ethene.
Diborane	Diborane is the chemical compound consisting of boron and hydrogen with the formula B_2H_6. It is a colorless gas at room temperature with a repulsively sweet odor. Diborane mixes well with air, easily forming explosive mixtures.
Diol	A Diol or glycol is a chemical compound containing two hydroxyl groups (-OH groups) Vicinal Diol s have hydroxyl groups attached to adjacent atoms. Examples of vicinal Diol compounds are ethylene glycol and propylene glycol. Geminal Diol s have hydroxyl groups bonded to the same atom.

Electrophile	In chemistry, an Electrophile (literally electron-lover) is a reagent attracted to electrons that participates in a chemical reaction by accepting an electron pair in order to bond to a nucleophile. Because Electrophile s accept electrons, they are Lewis acids Most Electrophile s are positively charged, have an atom which carries a partial positive charge, or have an atom which does not have an octet of electrons.
Reagent	A Reagent or reactant is a substance or compound consumed during a chemical reaction. Solvents and catalysts, although they are involved in the reaction, are usually not referred to as reactants. Although the terms reactant and Reagent are often used interchangeably, a Reagent is more specifically 'a test substance that is added to a system in order to bring about a reaction or to see whether a reaction occurs'.
Crown ethers	Crown ethers are heterocyclic chemical compounds that consist of a ring containing several ether groups. The most common Crown ethers are oligomers of ethylene oxide, the repeating unit being ethyleneoxy, i.e., $-CH_2CH_2O-$. Important members of this series are the tetramer (n = 4), the pentamer (n = 5), and the hexamer (n = 6.)
2-ethoxyethanol	2-Ethoxyethanol, also known by the trademark Cellosolve or ethyl cellosolve, is a solvent used widely in commercial and industrial applications. It is a clear, colorless, nearly odorless liquid that is miscible with water, ethanol, diethyl ether, acetone, and ethyl acetate. 2-Ethoxyethanol can be manufactured by the reaction of ethylene oxide with ethanol.
Anisole	Anisole is the organic compound with the formula $CH_3OC_6H_5$. This colorless liquid has a smell reminiscent of anise seed. It is used as a precursor to other organic compounds.
Chloromethyl methyl ether	Chloromethyl methyl ether is a compound with formula CH_3OCH_2Cl. It is a chloroalkyl ether. It is used as an alkylating agent and industrial solvent to manufacture dodecylbenzyl chloride, water repellents, ion-exchange resins, polymers, and as a chloromethylation reagent.
Alkanes	Alkanes are chemical compounds that consist only of the elements carbon and hydrogen, wherein these atoms are linked together exclusively by single bonds without any cyclic structure Alkanes belong to a homologous series of organic compounds in which the members differ by a constant relative atomic mass of 14. Each carbon atom must have 4 bonds, and each hydrogen atom must be joined to a carbon atom
Alkyl	In chemistry, an Alkyl group is a hydrocarbon; typically an Alkyl is a part of a larger molecule. The term is usually used loosely, there is no general formula for an Alkyl group. In structural formulae, an Alkyl group is represented with an R. Usually, Alkyl groups resemble hydrocarbons, but with one less hydrogen atom.
Cyclohexene	Cyclohexene is a colorless clear liquid cycloalkene with an intense aversive characteristic sharp smell reminiscent of an oil refinery.

It is not very stable upon long term storage with exposure to light and air and should be distilled before use to eliminate peroxides. A common experiment among beginning organic chemistry students is the acid catalyzed dehydration of cyclohexanol with distillative removal of the resulting Cyclohexene from the reaction mixture:

- Refractive Index: 1.4465
- Critical temperature: 287.2 °C (560.4 K)

Epoxide

An Epoxide is a cyclic ether with three ring atoms. This ring approximately defines an equilateral triangle, which makes it highly strained. The strained ring makes Epoxide s more reactive than other ethers.

Ethylene

Ethylene is the chemical compound with the formula C_2H_4. It is the simplest alkene. Because it contains a carbon-carbon double bond, Ethylene is called an unsaturated hydrocarbon or an olefin.

Ethylene oxide

Ethylene oxide, also called oxirane, is the organic compound with the formula C_2H_4O. This colorless flammable gas with a faintly sweet odor is the simplest epoxide, a three-membered ring consisting of two carbon and one oxygen atom. It is commonly handled and shipped as a refrigerated liquid. It is the chief precursor to ethylene glycol and other high-volume chemicals but is also used for medical sterilization.

Heteroatom

In the nomenclature of organic chemistry, a Heteroatom is any atom that is not carbon or hydrogen. It is typically, but not exclusively, nitrogen, oxygen, sulfur, phosphorus, boron, chlorine, bromine, or iodine.

In the description of protein structure, particularly in the now-deprecated Protein Data Bank file format, a Heteroatom record (HETATM) describes an atom belonging to a small molecule cofactor rather than to part of a biopolymer chain.

Heterocyclic compounds

Heterocyclic compounds are organic compounds containing at least one atom of carbon, and at least one element other than carbon, such as sulfur, oxygen or nitrogen within a ring structure. These structures may comprise either simple aromatic rings or non-aromatic rings. Some examples are pyridine (C_5H_5N), pyrimidine ($C_4H_4N_2$) and dioxane ($C_4H_8O_2$.)

Methylcyclohexane

Methylcyclohexane is a colourless liquid with a faint benzene-like odour. Its molecular formula is C_7H_{14}. Methylcyclohexane is used in organic synthesis and as a solvent for cellulose ethers.

Isobutylene

Isobutylene is a hydrocarbon of significant industrial importance. It is a four-carbon branched alkene (olefin), one of the four isomers of butylene. At standard temperature and pressure it is a colorless flammable gas.

Dioxin

Dioxin is a heterocyclic, organic, antiaromatic compound with the chemical formula $C_4H_4O_2$. There are two isomers, 1,2-Dioxin and 1,4-Dioxin (or p-Dioxin.) Their chemical structures are shown below.

Ethyl carbamate

Ethyl carbamate is a substance first prepared in the nineteenth century. Structurally it is an ester of carbamic acid, i. e., Ethyl carbamate as shown.

Furan

Furan is a heterocyclic organic compound. It is typically derived by the thermal decomposition of pentose-containing materials, cellulosic solids especially pine-wood. Furan is a colorless, flammable, highly volatile liquid with a boiling point close to room temperature.

Methylcyclopropane

Methylcyclopropane is the organic compound with the chemical formula $C_3H_5CH_3$. This colourless gas is the mono methyl derivative of cyclopropane.

Methylcyclopropane, like many cyclopropanes, undergoes ring-opening reactions.

Tetrahydropyran

Tetrahydropyran is the organic compound consisting of a saturated six-membered ring containing five carbon atoms and one oxygen atom.

The Tetrahydropyran ring system is the core of pyranose sugars.

One classic procedure for the organic synthesis of Tetrahydropyran is by hydrogenation with Raney nickel of dihydropyran.

Oxetane

Oxetane is the heterocyclic organic compound having a four-membered ring with three carbon atoms and one oxygen atom with the molecular formula C_3H_6O. The term Oxetane may also refer more generally to any organic compound containing an Oxetane ring.

A typical well known method of preparation is the reaction of potassium hydroxide with 3-chloropropyl acetate at 150°C Yield of Oxetane made this way is ca. 40%, as the synthesis can obviously lead to a variety of by-products.

Pyran

In chemistry, a Pyran is a six membered heterocyclic ring consisting of five carbon atoms and one oxygen atom and containing two double bonds. The molecular formula is C_5H_6O. There are two isomers of Pyran that differ by the location of the double bonds. In 2H-Pyran, the saturated carbon is at position 2, whereas in 4H-Pyran, the saturated carbon is at position 4.

Alpha carbon

The alpha carbon in organic chemistry refers to the first carbon that attaches to a functional group (the carbon is attached at the first position.) By extension, the second carbon is the beta carbon, and so on. This nomenclature can also be applied to the hydrogen atoms attached to the carbons.

Alpha cleavage

Alpha cleavage, (α-cleavage) in organic chemistry, refers to the act of breaking the carbon-carbon bond, adjacent to the carbon bearing a specified functional group.

Generally this topic is discussed when covering mass spectrometry and occurs generally by the same mechanisms.

For example of a mechanism of Alpha cleavage, an electron is knocked off an atom (usually by electron collision) to form a radical cation.

Alkoxide

An Alkoxide is the conjugate base of an alcohol and therefore consists of an organic group bonded to a negatively charged oxygen atom. They can be written as RO^-, where R is the organic substituent. Alkoxide s are strong bases and, when R is not bulky, good nucleophiles and good ligands.

Williamson ether synthesis	The Williamson ether synthesis is an organic reaction, forming an ether from a organohalide and an alcohol. This reaction was developed by Alexander Williamson in 1850 . Typically it involves the reaction of an alkoxide ion with a primary alkyl halide via an S_N2 reaction.
Oxalate	An Oxalate is the deprotonated, charged form of oxalic acid or an ester of oxalic acid. As a salt, the Oxalate anion has the chemical formula $C_2O_4^{2-}$ or $(COO)_2^{2-}$. Consumption of oxalates (for example, the grazing of animals on Oxalate-containing plants such as greasewood), or human consumption of Sorrel may result in kidney disease or even death due to Oxalate poisoning.
Phenol	In organic chemistry, phenols, sometimes called phenolics, are a class of chemical compounds consisting of a hydroxyl group (-OH) bonded directly to an aromatic hydrocarbon group. The simplest of the class is Phenol Phenol - the simplest of the phenols. Although similar to alcohols, phenols have unique properties and are not classified as alcohols (since the hydroxyl group is not bonded to a saturated carbon atom.)
Ethanol	Ethanol pure alcohol, grain alcohol is a volatile, flammable, colorless liquid. It is a psychoactive drug, best known as the type of alcohol found in alcoholic beverages and in modern thermometers. Ethanol is one of the oldest recreational drugs.
Methanol	Methanol, also known as methyl alcohol, carbinol, wood alcohol, wood naphtha or wood spirits, is a toxic chemical with chemical formula CH_3OH Drinking even small amounts can cause blindness. It is the simplest alcohol, and is a light, volatile, colourless, flammable, toxic liquid with a distinctive odor that is very similar to but slightly sweeter than ethanol (drinking alcohol.)
Norepinephrine	Noradrenaline (BAN) or Norepinephrine is a catecholamine with dual roles as a hormone and a neurotransmitter. As a stress hormone, Norepinephrine affects parts of the brain where attention and responding actions are controlled. Along with epinephrine, Norepinephrine also underlies the fight-or-flight response, directly increasing heart rate, triggering the release of glucose from energy stores, and increasing blood flow to skeletal muscle.
Ethyl phenyl ether	Ethyl phenyl ether or phenetole is an organic compound that is an ether. Ethyl phenyl ether has the same properties as some other ethers, such as volatility, explosive vapors, and the ability to form peroxides. Its boiling point was measured through experimental work at Karolinska Institutet, Stockholm, Sweden, at the institution for Medical Biochemistry and Biophysics.
Autoxidation	Autoxidation is any oxidation that occurs in open air or in presence of oxygen and/or UV radiation and forms peroxides and hydroperoxides. A classic example of Autoxidation is that of simple ethers like diethyl ether, whose peroxides can be dangerously explosive. It can be considered to be a slow, flameless combustion of materials by reaction with oxygen.

Dimethyl sulfide	Dimethyl sulfide or methylthiomethane is an organosulfur compound with the formula $(CH_3)_2S$. Dimethyl sulfide is a water-insoluble flammable liquid that boils at 37°C and has a characteristic disagreeable odor. It is a component of the smell produced from cooking of certain vegetables, notably corn, cabbage, beetroot and seafoods. It is also an indication of bacterial infection in malt production and brewing.
Sodium	Sodium is a metallic element with a symbol Na and atomic number 11. It is a soft, silvery-white, highly reactive metal and is a member of the alkali metals within 'group 1' (formerly known as 'group IA'.) It has only one stable isotope, ^{23}Na.
Sulfone	A Sulfone is a chemical compound containing a sulfonyl functional group attached to two carbon atoms. The central sulfur atom is twice double bonded to oxygen and has two further hydrocarbon substituents. The general structural formula is R-S(=O)(=O)-R' where R and R' are the organic groups.
Sulfoxide	A Sulfoxide is a chemical compound containing a sulfinyl functional group attached to two carbon atoms. Sulfoxides can be considered as oxidized sulfides. (The use of the alternative spelling sulphoxide is discouraged by IUPAC.)
Thioether	A Thioether is a functional group in organic chemistry that has the structure R^1-S-R^2 as shown on right. Like many other sulfur-containing compounds, volatile thioethers characteristically have foul odors. A Thioether is similar to an ether except that it contains a sulfur atom in place of the oxygen.
Propyl	In organic chemistry, Propyl is a three-carbon alkyl substituent with chemical formula $-C_3H_7$. It is the substituent form of the alkane propane. For example: Propyl ethanoate, also called Propyl acetate. This is Propyl ethanoate, an ester.
Sulfonium	A Sulfonium ion, also known as sulphonium ion and sulfanium ion, is a positively charged sulfur ion carrying three alkyl groups as substituents (S^+R_3.) Ionic compounds consisting of a positively charged Sulfonium cation and a negatively charged anion are called Sulfonium salts. Sulfonium compounds can be synthesized from the reaction of dialkylsulfides with alkyl halides: $CH_3\text{-}S\text{-}CH_3 + CH_3\text{-}I \rightarrow (CH_3)_3S^+ I^-$ (trimethylsulfonium iodide) In the above reaction, the alkylation of a sulfide, the sulfur atom has two lone electron pairs.
Epinephrine	Epinephrine is a hormone and neurotransmitter that participates in the 'fight or flight' response of the sympathetic nervous system. It is a catecholamine, a sympathomimetic monoamine produced by the adrenal glands from the amino acids phenylalanine and tyrosine.

The term Epinephrine is derived from the Greek roots epi- and nephros, and literally means on the kidney, in reference to the gland's anatomic location.

Concerted reaction

In chemistry, a Concerted reaction is a chemical reaction in which all bond breaking and bond making occurs in a single step. Reactive intermediates or other unstable high energy intermediates are not involved. Concerted reaction rates tend not to depend on solvent polarity ruling out large buildup of charge in the transition state.

Halohydrin

A Halohydrin is a type of chemical compound or functional group in which one carbon atom has a halogen substituent, and an adjacent carbon atom has a hydroxyl substituent.

Halohydrin formation:

- from an alkene in a Halohydrin formation reaction
- from an epoxide by a hydrohalic acid

Halohydrin reactions:

- In presence of a base, such as potassium hydroxide, a Halohydrin may undergo internal S_N2 reaction to form an epoxide. This is the reverse of the formation reaction from an epoxide.
- Epoxidation in biological systems can be catalyzed by Halohydrin dehalogenase.

.

Amide

In chemistry, an Amide is one of three kinds of compounds:

- (sometimes called acid Amide the organic functional group characterized by a carbonyl group (C=O) linked to a nitrogen atom (N), or a compound that contains this functional group (pictured to the right); or
- a kind of anion or
- any organic compound derived by the replacement of a hydroxyl group by an amino group.

Amide s are the most stable of all the carbonyl functional groups.

Many chemists make a pronunciation distinction between the two, saying /É™Ë̈mïË□d/ for the carbonyl-nitrogen compound and /ËˆæmaÉªd/ for the anion. Others substitute one of these with , while still others pronounce both , making them homonyms.

In the first sense referred to above, an Amide is an amine where one of the nitrogen substituents is an acyl group; it is generally represented by the formula: $R_1(CO)NR_2R_3$, where either or both R_2 and R_3 may be hydrogen.

2-butene

2-Butene is a compound with formula C_4H_8. It is one of the isomers of butene. It is a good example of cis-trans isomerism.

Steroid

A Steroid is a terpenoid lipid characterized by a carbon skeleton with four fused rings, generally arranged in a 6-6-6-5 fashion. Steroids vary by the functional groups attached to these rings and the oxidation state of the rings. Hundreds of distinct steroids are found in plants, animals, and fungi.

Amine

Amine s are organic compounds and functional groups that contain a basic nitrogen atom with a lone pair. Amine s are derivatives of ammonia, wherein one or more hydrogen atoms have been replaced by a substituent such as an alkyl or aryl group. Important Amine s include amino acids, biogenic Amine s, trimethyl Amine , and aniline; see Category: Amine s for a list of Amine s.

Cyclopentane

Cyclopentane is a highly flammable alicyclic hydrocarbon with chemical formula C_5H_{10} and CAS number 287-92-3, consisting of a ring of five carbon atoms each bonded with two hydrogen atoms above and below the plane. It occurs as a colorless liquid with a petrol-like odor. Its melting point is −94 °C and its boiling point is 49 °C. The typical structure of Cyclopentane is the 'envelope' conformation.

Cyclopentane is used in the manufacture of synthetic resins and rubber adhesives and also as a blowing agent in the manufacture of polyurethane insulating foam, as found in may domestic appliances such as refrigerators and freezers, replacing environmentally damaging alternatives such as CFC-11 and HCFC-141b

More advanced technologies, such as computer hard drives and outerspace equipment employ multiply-alkylated Cyclopentane lubricants because of their extremely low volatility.

Cyclopentene

Cyclopentene is a chemical compound with the formula C_5H_8. It is a colorless liquid with a petrol-like odor. It is one of the cycloalkenes.

Ethanolamine

Ethanolamine is an organic chemical compound that is both a primary amine and a primary alcohol Like other amines, mono Ethanolamine acts as a weak base. Ethanolamine is a toxic, flammable, corrosive, colorless, viscous liquid with an odor similar to that of ammonia.

Ethylmagnesium bromide

Ethylmagnesium bromide is an organomagnesium compound. Apart from acting as the synthetic equivalent of an ethyl anion synthon for nucleophilic addition, it may be used as a strong base to deprotonate various substrates such as alkynes:

$$RC{\equiv}CH + EtMgBr \rightarrow RC{\equiv}CMgBr + EtH$$

This application has been supplanted by the wide availability of organolithium reagents.

Ethylmagnesium bromide is commercially available, usually as a solution in diethyl ether or tetrahydrofuran.

Grignard reaction

The Grignard reaction is an organometallic chemical reaction in which alkyl- or aryl-magnesium halides , act as nucleophiles, attack electrophilic carbon atoms that are present within polar bonds to yield a carbon-carbon bond (compare to Wittig reaction), thus altering hybridization about the reaction center. The Grignard reaction is an important tool in the formation of carbon-carbon bonds and for the formation of carbon-phosphorus, carbon-tin, carbon-silicon, carbon-boron and other carbon-heteroatom bonds.

	The addition to the nucleophile is irreversible due to the high pK_a value of the alkyl component (pK_a = ~45.)
Organolithium reagent	An Organolithium reagent is an organometallic compound with a direct bond between a carbon and a lithium atom. As the electropositive nature of lithium puts most of the charge density of the bond on the carbon atom, effectively creating a carbanion, organolithium compounds are extremely powerful bases and nucleophiles. Organolithium reagents are industrially prepared by the reaction of an halocarbon with lithium metal, i.e. R-X + 2 Li → R-Li + LiX. A side reaction of this synthesis, especially with alkyl iodides, is the Wurtz reaction, in which an R-Li species reacts with an R-X species forming an R-R coupled product.
Epoxy	Epoxy or polyepoxide is a thermosetting polymer formed from reaction of an epoxide 'resin' with polyamine 'hardener'. Epoxy has a wide range of applications, including fiber-reinforced plastic materials and general purpose adhesives. Structure of unmodified Epoxy prepolymer.
Bisphenol A	Bisphenol A, commonly abbreviated as BPA, is an organic compound with two phenol functional groups. It is a difunctional building block of several important plastics and plastic additives. With an annual production of 2-3 million metric tonnes, it is an important monomer in the production of polycarbonate.
Epichlorohydrin	Epichlorohydrin is an organochlorine compound and an epoxide. This is a colorless liquid with a pungent, garlic-like odor, insoluble in water, but miscible with most polar organic solvents. Epichlorohydrin is a highly reactive compound and is used in the production of glycerol, plastics, and elastomers.

3-pentanone	3-Pentanone is a colorless liquid ketone with an odor like that of acetone. Its formula is $C_5H_{10}O$. It is soluble in about 25 parts water, and miscible with ethanol and diethyl ether. Two other ketones, which are isomers of 3-Pentanone, are 2-pentanone and methyl isopropyl ketone.
Diels-Alder reaction	The Diels-Alder reaction is an organic chemical reaction (specifically, a cycloaddition) between a conjugated diene and a substituted alkene, commonly termed the dienophile, to form a substituted cyclohexene system. The reaction can proceed even if some of the atoms in the newly-formed ring are not carbon. Some of the Diels-Alder reaction s are reversible; the decomposition reaction of the cyclic system is then called the Retro-Diels-Alder.
Hydrogenation	Hydrogenation is the chemical reaction that results from the addition of hydrogen (H_2.) The process is usually employed to reduce or saturate organic compounds. The process typically constitutes the addition of pairs of hydrogen atoms to a molecule.
Allene	An Allene is a hydrocarbon in which one atom of carbon is connected by double bonds with two other atoms of carbon. Allene also is the common name for the parent compound of this series, propadiene. Such pair of bonds make Allene s much more reactive than other alkenes.
Enol	Enol s are alkenes with a hydroxyl group affixed to one of the carbon atoms composing the double bond. Enol s and carbonyl compounds are in fact isomers; this is called keto Enol tautomerism: The Enol form is shown above. It is usually unstable, does not survive long, and changes into the keto form shown on the right.
Ethanol	Ethanol pure alcohol, grain alcohol is a volatile, flammable, colorless liquid. It is a psychoactive drug, best known as the type of alcohol found in alcoholic beverages and in modern thermometers. Ethanol is one of the oldest recreational drugs.
Conjugated system	A Conjugated system occurs in an organic compound where atoms covalently bond with alternating single and multiple (e.g. double) bonds (e.g., C=C-C=C-C) and influence each other to produce a region called electron delocalization. In this region electrons do not belong to a single bond or atom, but rather a group. For example, phenol (C_6H_5OH, benzene with hydroxyl group) (diagramatically has alternating single and double bonds), which has a system of 6 electrons above and below the flat planar ring, as well as around the hydroxyl group.
Ethylene	Ethylene is the chemical compound with the formula C_2H_4. It is the simplest alkene. Because it contains a carbon-carbon double bond, Ethylene is called an unsaturated hydrocarbon or an olefin.
Schiff base	A Schiff base is a functional group that contains a carbon-nitrogen double bond with the nitrogen atom connected to an aryl or alkyl group--but not hydrogen. Schiff bases are of the general formula $R_1R_2C=N-R_3$, where R_3 is an aryl or alkyl group that makes the Schiff base a stable imine. A Schiff base derived from an aniline, where R_3 is a phenyl or substituted phenyl, can be called an anil.

Configuration	The configuration of a molecule is the permanent geometry that results from the spatial arrangement of its bonds. The ability of the same set of atoms to form two or more molecules with different configurations is stereoisomerism. configuration is distinct from chemical conformation, a shape attainable by bond rotations.
Allyl	An Allyl group is an alkene hydrocarbon group with the formula $H_2C{=}CH{-}CH_2{-}$. It is made up of a vinyl group, $CH_2{=}CH{-}$, attached to a methylene $-CH_2$. For example Allyl alcohol has the structure $H_2C{=}CH{-}CH_2OH$. Another example of a simple Allyl compound is Allyl chloride.
Allyl alcohol	Allyl alcohol or 2-propen-1-ol is an organic compound with the formula $CH_2{=}CHCH_2OH$. It is a water soluble, colourless liquid with an ethanol like odour at low concentrations and a mustard-like pungent odour at higher concentration. Allyl alcohol is used as a pesticide and as a raw material for the production of many chemical compounds. Allyl alcohol can be obtained by many methods: hydrolysis of allyl chloride, by oxidation of propylene oxide with potassium alum at high temperature, dehydrogenation of propanol, and by the reaction of glycerol and formic acid.
Allyl bromide	Allyl bromide is an organic halide. Its refractive index is 1.4697 (20 °C, 589 nm.) Allyl bromide is an alkylating agent used in synthesis of polymers, pharmaceuticals, allyls and other organic compounds.
Radicals	In chemistry, Radicals are atoms, molecules, or ions with unpaired electrons on an otherwise open shell configuration. These unpaired electrons are usually highly reactive, so Radicals are likely to take part in chemical reactions. Radicals play an important role in combustion, atmospheric chemistry, polymerization, plasma chemistry, biochemistry, and many other chemical processes, including human physiology.
Electrophilic addition	In organic chemistry, an Electrophilic addition reaction is an addition reaction where, in a chemical compound, a pi bond is removed by the creation of two new covalent bonds. The substrate of an Electrophilic addition reaction must have a double bond or triple bond . The driving force for this reaction is the formation of an electrophile X^+ that forms a covalent bond with an electron-rich unsaturated C=C bond.
Alkynes	Alkynes are hydrocarbons that have a triple bond between two carbon atoms, with the formula C_nH_{2n-2}. Alkynes are traditionally known as acetylenes, although the name acetylene also refers specifically to C_2H_2, known formally (but rarely) as ethyne using IUPAC nomenclature. Like other hydrocarbons, Alkynes are generally hydrophobic but tend to be more reactive.
Cyclohexene	Cyclohexene is a colorless clear liquid cycloalkene with an intense aversive characteristic sharp smell reminiscent of an oil refinery. It is not very stable upon long term storage with exposure to light and air and should be distilled before use to eliminate peroxides. A common experiment among beginning organic chemistry students is the acid catalyzed dehydration of cyclohexanol with distillative removal of the resulting Cyclohexene from the reaction mixture: • Refractive Index: 1.4465 • Critical temperature: 287.2 °C (560.4 K)

N-bromosuccinimide

N-Bromosuccinimide or NBS is a chemical reagent which is used in radical substitution and electrophilic addition reactions in organic chemistry. NBS can be considered a convenient source of cationic bromine.

NBS will react with alkenes 1 in aqueous solvents to give bromohydrins 2.

Succinimide

Succinimide is a cyclic imide with the formula $C_4H_5NO_2$. It appears as a white or colorless crystalline solid, and has a faint odor of sweat. It is used in a variety of organic synthesis, as well as in some industrial silver plating processes.

Norepinephrine

Noradrenaline (BAN) or Norepinephrine is a catecholamine with dual roles as a hormone and a neurotransmitter.

As a stress hormone, Norepinephrine affects parts of the brain where attention and responding actions are controlled. Along with epinephrine, Norepinephrine also underlies the fight-or-flight response, directly increasing heart rate, triggering the release of glucose from energy stores, and increasing blood flow to skeletal muscle.

Halide

A Halide is a binary compound, of which one part is a halogen atom and the other part is an element or radical that is less electronegative than the halogen, to make a fluoride, chloride, bromide, iodide, or astatide compound. Many salts are Halide s. All Group 1 metals form Halide s with the halogens and they are white solids.

N-propyl bromide

N-Propyl bromide is a colorless, flammable chemical compound. It has the chemical formula C_3H_7Br. It is an organic solvent used for the cleaning of metal surfaces, removal of soldering residues from electronic circuit boards, and as an adhesive solvent.

2-heptanone

2-Heptanone, or methyl n-amyl ketone, is a ketone with the molecular formula $C_7H_{14}O$. It is a colorless, water-white liquid with a banana-like, fruity odor.

2-Heptanone is listed by the FDA as a 'food additive permitted for direct addition to food for human consumption' (21 CFR 172.515), and it occurs naturally in certain foods (e.g., beer, white bread, butter, various cheeses and potato chips.)

The mechanism of action of 2-Heptanone as a pheromone at odorant receptors in mice has been investigated.

Reaction mechanism

In chemistry, a Reaction mechanism is the step by step sequence of elementary reactions by which overall chemical change occurs .

Although only the net chemical change is directly observable for most chemical reactions, experiments can often be designed that suggest the possible sequence of steps in a Reaction mechanism.

A mechanism describes in detail exactly what takes place at each stage of a chemical transformation.

Cyclopentadiene

Cyclopentadiene is a chemical compound with the formula C_5H_6. This colorless liquid organic chemical has a strong and unpleasant odor. At room temperature, this cyclic diene dimerizes over the course of hours to give di Cyclopentadiene via a Diels-Alder reaction.

Alkanes	Alkanes are chemical compounds that consist only of the elements carbon and hydrogen, wherein these atoms are linked together exclusively by single bonds without any cyclic structure Alkanes belong to a homologous series of organic compounds in which the members differ by a constant relative atomic mass of 14. Each carbon atom must have 4 bonds, and each hydrogen atom must be joined to a carbon atom
Cycloaddition	A Cycloaddition is a pericyclic chemical reaction, in which 'two or more unsaturated molecules (or parts of the same molecule) combine with the formation of a cyclic adduct in which there is a net reduction of the bond multiplicity.' The resulting reaction is a cyclization reaction. Cycloaddition s are usually described by the backbone size of the participants. This would make the Diels-Alder reaction a [4 + 2 Cycloaddition and the 1,3-dipolar Cycloaddition a [3 + 2 Cycloaddition
Reagent	A Reagent or reactant is a substance or compound consumed during a chemical reaction. Solvents and catalysts, although they are involved in the reaction, are usually not referred to as reactants. Although the terms reactant and Reagent are often used interchangeably, a Reagent is more specifically 'a test substance that is added to a system in order to bring about a reaction or to see whether a reaction occurs'.
Pericyclic reaction	In organic chemistry, a Pericyclic reaction is a type of organic reaction wherein the transition state of the molecule has a cyclic geometry, and the reaction progresses in a concerted fashion. Pericyclic reactions are usually rearrangement reactions. The major classes of pericyclic reactions are: • Electrocyclic reactions • Cycloadditions • Sigmatropic reactions • Group transfer reactions • Cheletropic reactions In general, these are considered to be equilibrium processes, although it is possible to push the reaction in one direction by designing a reaction by which the product is at a significantly lower energy level; this is due to a unimolecular interpretation of Le Chatelier's principle. Pericyclic reactions often have related stepwise radical processes associated with them.
1-hexanol	1-Hexanol is an organic alcohol with a six carbon chain and a condensed structural formula of $CH_3(CH_2)_5OH$. This colorless liquid is slightly soluble in water, but miscible with ether and ethanol. Two additional straight chain isomers of 1-Hexanol exist, 2-hexanol and 3-hexanol, both of which differ by the location of the hydroxyl group. Many isomeric alcohols have the formula $C_6H_{13}OH$. 1-Hexanol is believed to be a component of the odour of freshly mowed grass.
Hexane	Hexane is an alkane hydrocarbon with the chemical formula $CH_3(CH_2)_4CH_3$ or C_6H_{14}. The 'hex' prefix refers to its six carbons, while the 'ane' ending indicates that its carbons are connected by single bonds. Hexane isomers are largely unreactive, and are frequently used as an inert solvent in organic reactions because they are very non-polar.

Hexanoic acid	Hexanoic acid, is the carboxylic acid derived from hexane with the general formula $C_5H_{11}COOH$. It is a colorless oily liquid with an odor reminiscent of goats or other barnyard animals. It is a fatty acid found naturally in various animal fats and oils, and is one of the chemicals that gives the decomposing fleshy seed coat of the ginkgo its characteristic unpleasant odor. The salts and esters of this acid are known as hexanoates or caproates.

1,2-dichlorobenzene	1,2-Dichlorobenzene, or ortho-dichlorobenzene, is an organic compound with the formula $C_6H_4Cl_2$. This colourless liquid is poorly soluble in water but miscible with most organic solvents. It is a derivative of benzene, consisting of two adjacent chlorine centers.
Aliphatic compounds	In organic chemistry, compounds composed of carbon and hydrogen are divided into two classes: aromatic compounds, which contain benzene rings or similar rings of atoms, and Aliphatic compounds, which do not contain aromatic rings.Aliphatic compounds can be cyclic, like cyclohexane, or acyclic, like hexane. They also can be saturated, like hexane, or unsaturated, like hexene. In Aliphatic compounds, carbon atoms can be joined together in straight chains, branched chains, or non-aromatic rings (in which case they are called alicyclic.)
Benzene	Benzene, or benzol, is an organic chemical compound and a known carcinogen with the molecular formula C_6H_6. It is sometimes abbreviated Ph-H. Benzene is a colorless and highly flammable liquid with a sweet smell and a relatively high melting point. Because it is a known carcinogen, its use as an additive in gasoline is now limited, but it is an important industrial solvent and precursor in the production of drugs, plastics, synthetic rubber, and dyes.
Alcohol	In chemistry, an Alcohol is any organic compound in which a hydroxyl group (-OH) is bound to a carbon atom of an alkyl or substituted alkyl group. The general formula for a simple acyclic Alcohol is $C_nH_{2n+1}OH$. In common terms, the word Alcohol refers to ethanol, the type of Alcohol found in Alcohol ic beverages. Ethanol is a colorless, volatile liquid with a mild odor which can be obtained by the fermentation of sugars.
Potassium permanganate	Potassium permanganate is the inorganic chemical compound $KMnO_4$, a water soluble salt consisting of equal mole amounts of potassium (K^+) and permanganate (MnO_4^-, officially called manganate (VII)) ions. This salt, formerly known as permanganate of potash or Condy's crystals is a strong oxidizing agent. It dissolves in water to give deep purple solutions, evaporation of which gives prismatic purplish-black glistening crystals.
Alkynes	Alkynes are hydrocarbons that have a triple bond between two carbon atoms, with the formula C_nH_{2n-2}. Alkynes are traditionally known as acetylenes, although the name acetylene also refers specifically to C_2H_2, known formally (but rarely) as ethyne using IUPAC nomenclature. Like other hydrocarbons, Alkynes are generally hydrophobic but tend to be more reactive.
Cyclohexene	Cyclohexene is a colorless clear liquid cycloalkene with an intense aversive characteristic sharp smell reminiscent of an oil refinery. It is not very stable upon long term storage with exposure to light and air and should be distilled before use to eliminate peroxides. A common experiment among beginning organic chemistry students is the acid catalyzed dehydration of cyclohexanol with distillative removal of the resulting Cyclohexene from the reaction mixture: • Refractive Index: 1.4465 • Critical temperature: 287.2 °C (560.4 K)

Propane	Propane is a three-carbon alkane, normally a gas, but compressible to a transportable liquid. It is derived from other petroleum products during oil or natural gas processing. It is commonly used as a fuel for engines, oxy-gas torches, barbecues, portable stoves and residential central heating.
Annulene	Annulene s are completely conjugated monocyclic hydrocarbons. They have the general formula C_nH_n (when n is an even number) or C_nH_{n+1} (when n is an odd number.) The IUPAC naming conventions are that Annulene s with 7 or more carbon atoms are named as [n Annulene where n is the number of carbon atoms in their ring, though sometimes the smaller Annulene s are referred to using the same notation, and benzene is sometimes referred to simply as Annulene
Cyclobutadiene	Cyclobutadiene is the smallest [n]-annulene (-annulene), an extremely unstable hydrocarbon having a lifetime shorter than five seconds in the free state. It has chemical formula C_4H_4 and a rectangular structure verified by infrared studies. This is in contrast to the square geometry predicted by simple Hückel theory.
Cyclodecapentaene	Cyclodecapentaene or annulene is an annulene with molecular formula $C_{10}H_{10}$. This organic compound is a conjugated 10 pi electron cyclic system and according to Huckel's rule it should display aromaticity. It is not aromatic, however, because of a combination of steric strain and angular strain.
Configuration	The configuration of a molecule is the permanent geometry that results from the spatial arrangement of its bonds. The ability of the same set of atoms to form two or more molecules with different configurations is stereoisomerism. configuration is distinct from chemical conformation, a shape attainable by bond rotations.
Naphthalene	Naphthalene, also known as naphthalin, naphthaline, tar camphor, white tar, albocarbon, or antimite and not to be confused with naphtha, is a crystalline, aromatic, white, solid hydrocarbon with formula $C_{10}H_8$ and the structure of two fused benzene rings. It is best known as the traditional, primary ingredient of mothballs. It is volatile, forming an inflammable vapor, and readily sublimes at room temperature, producing a characteristic odor that is detectable at concentrations as low as 0.08 ppm by mass.
Cyclopentadiene	Cyclopentadiene is a chemical compound with the formula C_5H_6. This colorless liquid organic chemical has a strong and unpleasant odor. At room temperature, this cyclic diene dimerizes over the course of hours to give di Cyclopentadiene via a Diels-Alder reaction.
Pyridine	Pyridine is a simple aromatic heterocyclic organic compound with the chemical formula C_5H_5N used as a precursor to agrochemicals and pharmaceuticals, and is also an important solvent and reagent. It is structurally related to benzene, wherein one CH group in the aromatic six-membered ring is replaced by a nitrogen atom. It exists as a colorless liquid with a distinctive, unpleasant fish-like odor.
Pyrrole	Pyrrole is a heterocyclic aromatic organic compound, a five-membered ring with the formula C_4H_4NH. Substituted derivatives are also called pyrroles. For example, $C_4H_4NCH_3$ is N-methylpyrrole. Porphobilinogen is a trisubstituted Pyrrole, which is the biosynthetic precursor to many natural products.
Porphobilinogen	Porphobilinogen is a pyrrole involved in porphyrin metabolism. It is generated by aminolevulinate (ALA) and the enzyme ALA dehydratase. PBG is then converted into hydroxymethyl bilane by the enzyme Porphobilinogen deaminase.

Pyrimidine

Pyrimidine is a heterocyclic aromatic organic compound similar to benzene and pyridine, containing two nitrogen atoms at positions 1 and 3 of the six-member ring. It is isomeric with two other forms of diazine.

Three nucleobases found in nucleic acids (cytosine, thymine, and uracil) are Pyrimidine derivatives:

In DNA and RNA, these bases form hydrogen bonds with their complementary purines.

Epoxide

An Epoxide is a cyclic ether with three ring atoms. This ring approximately defines an equilateral triangle, which makes it highly strained. The strained ring makes Epoxide s more reactive than other ethers.

Furan

Furan is a heterocyclic organic compound. It is typically derived by the thermal decomposition of pentose-containing materials, cellulosic solids especially pine-wood. Furan is a colorless, flammable, highly volatile liquid with a boiling point close to room temperature.

Imidazole

Imidazole is an organic compound with the formula $C_3H_4N_2$. This aromatic heterocyclic is classified as an alkaloid. Imidazole refers to the parent compound whereas Imidazole s are a class of heterocycles with similar ring structure but varying substituents.

Purine

Purine is a heterocyclic aromatic organic compound, consisting of a pyrimidine ring fused to an imidazole ring. Purines, including substituted purines and their tautomers, are the most widely distributed kind of nitrogen-containing heterocycle in nature.

Purines and pyrimidines make up the two groups of nitrogenous bases, including the two groups of nucleotide bases.

Thiophene

Thiophene is the heterocyclic compound with the formula C_4H_4S. Consisting of a flat five-membered ring, it is aromatic as indicated by its extensive substitution reactions. Related to Thiophene are benzothiophene and dibenzothiophene, containing the Thiophene ring fused with one and two benzene rings, respectively. Compounds analogous to Thiophene include furan (C_4H_4O) and pyrrole (C_4H_4NH.)

Anthracene

Anthracene is a solid polycyclic aromatic hydrocarbon consisting of three fused benzene rings derived from coal-tar or other residues of thermal pyrolysis. Anthracene is used in the artificial production of the red dye alizarin. It is also used in wood preservatives, insecticides, and coating materials.

Hydrocarbon

In organic chemistry, a Hydrocarbon is an organic compound consisting entirely of hydrogen and carbon. With relation to chemical terminology, aromatic Hydrocarbon s or arenes, alkanes, alkenes and alkyne-based compounds composed entirely of carbon or hydrogen are referred to as 'pure' Hydrocarbon s, whereas other Hydrocarbon s with bonded compounds or impurities of sulfur or nitrogen, are referred to as 'impure', and remain somewhat erroneously referred to as Hydrocarbon s.

Hydrocarbon s are referred to as consisting of a 'backbone' or 'skeleton' composed entirely of carbon and hydrogen and other bonded compounds, and have a functional group that generally facilitates combustion.

Phenanthrene

Phenanthrene is a polycyclic aromatic hydrocarbon composed of three fused benzene rings. The name Phenanthrene is a composite of phenyl and anthracene. It provides the framework for the steroids.

Aromatic Hydrocarbon	An Aromatic hydrocarbon or arene is a hydrocarbon, of which the molecular structure incorporates one or more planar sets of six carbon atoms that are connected by delocalised electrons numbering the same as if they consisted of alternating single and double covalent bonds. The term 'aromatic' was assigned before the physical mechanism determining aromaticity was discovered, and was derived from the fact that many of the compounds have a sweet scent. The configuration of six carbon atoms in aromatic compounds is known as a benzene ring, after the simplest possible such hydrocarbon, benzene.
Pyrene	Pyrene is a polycyclic aromatic hydrocarbon (PAH) consisting of four fused benzene rings, resulting in a flat aromatic system. This colourless solid is the smallest peri-fused polycyclic aromatic hydrocarbon - one where the rings are fused through more than one face. Pyrene forms during incomplete combustion of organic compounds.
Carbon	Carbon is the chemical element with symbol C and atomic number 6. As a member of group 14 on the periodic table, it is nonmetallic and tetravalent--making four electrons available to form covalent chemical bonds. There are three naturally occurring isotopes, with ^{12}C and ^{13}C being stable, while ^{14}C is radioactive, decaying with a half-life of about 5730 years.
Benzimidazole	Benzimidazole is a heterocyclic aromatic organic compound. This bicyclic compound consists of the fusion of benzene and imidazole. The most prominent Benzimidazole compound in nature is N-ribosyl-dimethyl Benzimidazole , which serves as an axial ligand for cobalt in vitamin B12.
Benzofuran	Benzofuran is the heterocyclic compound consisting of fused benzene and furan rings. It is the parent of many related compounds with more complex structures. For example, psoralen is a Benzofuran derivative that occurs in several plants.
Benzothiophene	Benzothiophene is an aromatic organic compound with a molecular formula C_8H_6S and an odor similar to naphthalene (mothballs.) It occurs naturally as a constituent of petroleum-related deposits such as lignite tar. Benzothiophene has no household use.
Indole	Indole is an aromatic heterocyclic organic compound. It has a bicyclic structure, consisting of a six-membered benzene ring fused to a five-membered nitrogen-containing pyrrole ring. The participation of the nitrogen lone electron pair in the aromatic ring means that Indole is not a base, and it does not behave like a simple amine.
Quinine	Quinine is a natural white crystalline alkaloid having antipyretic , antimalarial, analgesic (painkilling), and anti-inflammatory properties and a bitter taste. It is a stereoisomer of quinidine. Quinine was the first effective treatment for malaria caused by Plasmodium falciparum, appearing in therapeutics in the 17th century.
Quinoline	Quinoline 1-benzazine is a heterocyclic aromatic organic compound. It has the formula C_9H_7N and is a colourless hygroscopic liquid with a strong odour. Aged samples, if exposed to light, become yellow and later brown.
Acetone	Acetone is the organic compound with the formula $OC(CH_3)_2$. This colorless, mobile, flammable liquid is the simplest example of the ketones. Owing to the fact that Acetone is miscible with water, and it serves as an important solvent in its own right, typically the solvent of choice for cleaning purposes in the laboratory.

Acetophenone	Acetophenone is the organic compound with the formula $C_6H_5C(O)CH_3$. It is the simplest aromatic ketone. This colourless, viscous liquid is a precursor to useful resins and fragrances.
Aniline	Aniline, phenylamine or aminobenzene is an organic compound with the formula C_6H_7N. It is the simplest and one of the most important aromatic amines, being used as a precursor to more complex chemicals. Its main application is in the manufacture of polyurethane. Like most volatile amines, it possesses the somewhat unpleasant odour of rotten fish and also has a burning aromatic taste; it is a highly-acrid poison.
Anisole	Anisole is the organic compound with the formula $CH_3OC_6H_5$. This colorless liquid has a smell reminiscent of anise seed. It is used as a precursor to other organic compounds.
Benzaldehyde	Benzaldehyde is a chemical compound consisting of a benzene ring with an aldehyde substituent. It is the simplest representative of the aromatic aldehydes and one of the most industrially used members of this family of compounds. At room temperature it is a colorless liquid with a characteristic and pleasant almond-like odor: Benzaldehyde is an important component of the scent of almonds, hence its typical odor.
Benzenesulfonic acid	Benzenesulfonic acid is an organic compound with the formula $C_6H_5SO_3H$. It is the simplest aromatic sulfonic acid. It is a colourless deliquescent sheet crystal or white wax solid that is soluble in water and ethanol, slightly soluble in benzene and insoluble in carbon disulfide and diethyl ether. It is usually stored in the form of alkali metal salts.
Benzoic acid	Benzoic acid, $C_7H_6O_2$ (or C_6H_5COOH), is a colorless crystalline solid and the simplest aromatic carboxylic acid. The name derived from gum benzoin, which was for a long time the only source for Benzoic acid. This weak acid and its salts are used as a food preservative.
Butylated hydroxyanisole	Butylated hydroxyanisole is an antioxidant consisting of a mixture of two isomeric organic compounds, 2-tert-butyl-4-hydroxyanisole and 3-tert-butyl-4-hydroxyanisole. It is prepared from 4-methoxyphenol and isobutylene. It is a waxy solid used in certain amounts as a food additive with the E number E320.
Nitrobenzene	Nitrobenzene, also known as nitrobenzol or oil of mirbane, is an organic compound with the chemical formula $C_6H_5NO_2$. Nitrobenzene is a water-insoluble oil which exhibits a pale yellow to yellow-brown coloration in liquid form (at room temperature and pressure) with an almond-like odor. When frozen, it appears as a greenish-yellow crystal.
Phenol	In organic chemistry, phenols, sometimes called phenolics, are a class of chemical compounds consisting of a hydroxyl group (-OH) bonded directly to an aromatic hydrocarbon group. The simplest of the class is Phenol Phenol - the simplest of the phenols. Although similar to alcohols, phenols have unique properties and are not classified as alcohols (since the hydroxyl group is not bonded to a saturated carbon atom.)
Styrene	Styrene, also known as vinyl benzene as well as many other names , is an organic compound with the chemical formula $C_6H_5CH{=}CH_2$. This aromatic hydrocarbon is a colorless oily liquid that evaporates easily and has a sweet smell, although high concentrations confer a less pleasant odor. Styrene is the precursor to polystyrene and several copolymers.

Toluene

Toluene phenylmethane, and Toluol, is a clear water-insoluble liquid with the typical smell of paint thinners, redolent of the sweet smell of the related compound benzene. It is an aromatic hydrocarbon that is widely used as an industrial feedstock and as a solvent. Like other solvents, Toluene is also used as an inhalant drug for its intoxicating properties; however this causes severe neurological harm.

Addition reaction

An Addition reaction in organic chemistry, is in its simplest terms an organic reaction where two or more molecules combine to form a larger one.

There are two main types of polar Addition reaction s:

- Electrophilic addition
- Nucleophilic addition

Other non-polar Addition reaction s exists as well:

- Free radical addition

Addition reaction s are limited to chemical compounds that have multiply-bonded atoms:

- Molecules with carbon-carbon double bonds (alkenes) or triple bonds (alkynes)
- Molecules with carbon - hetero double bonds like C=O or C=N

An Addition reaction is the opposite of an elimination reaction. For instance the hydration reaction of an alkene and the dehydration of an alcohol are addition-elimination pairs.

In the related Addition-elimination reaction an Addition reaction is followed by an elimination reaction.

Isomer

In chemistry, Isomer s are compounds with the same molecular formula but different structural formula. Isomer s do not necessarily share similar properties unless they also have the same functional groups. This should not be confused with a nuclear Isomer which involves a nucleus at different states of excitement.

1,3,5-trinitrobenzene

1,3,5-Trinitrobenzene is an explosive chemical compound. Its use has been replaced by trinitrotoluene which requires a higher energy to detonate.

Aryl

In the context of organic molecules, Aryl refers to any functional group or substituent derived from a simple aromatic ring, may it be phenyl, thiophenyl, indolyl, etc . 'Aryl' is used for the sake of abbreviation or generalization.

A simple Aryl group is phenyl, C_6H_5; it is derived from benzene.

Benzyl

In organic chemistry, Benzyl is the term used to describe the substituent or molecular fragment possessing the structure $C_6H_5CH_2$-. The abbreviation 'Bn' is frequently used to denote Benzyl moieties in nomenclature and structural depictions of chemical compounds. For example, Benzyl alcohol can be represented as BnOH. This abbreviation is not to be confused with 'Bz', which is the abbreviation for the benzoyl group $C_6H_5C(O)$-.

Cracking

In petroleum geology and chemistry, Cracking is the process whereby complex organic molecules such as kerogens or heavy hydrocarbons are broken down into simpler molecules (e.g. light hydrocarbons) by the breaking of carbon-carbon bonds in the precursors. The rate of Cracking and the end products are strongly dependent on the temperature and presence of any catalysts. Cracking, also referred to as pyrolysis, is the breakdown of a large alkane into smaller, more useful alkanes and an alkene.

Dimethyl sulfoxide

Dimethyl sulfoxide is the chemical compound with the formula $(CH_3)_2SO$. It was first synthesized in 1866 by the Russian scientist Alexander Saytzeff, who reported his findings in a German chemistry journal in 1867. This colorless liquid is an important polar aprotic solvent that dissolves both polar and nonpolar compounds and is miscible in a wide range of organic solvents as well as water. It has a distinctive property of penetrating the skin very readily, so that one can taste it soon after it comes into contact with the skin.

Diphenyl ether

Diphenyl ether is the organic compound with the formula $O(C_6H_5)_2$. The molecule is subject to reactions typical of other phenyl rings, including hydroxylation, nitration, halogenation, sulfonation, and Friedel-Crafts alkylation or acylation. This simple diaryl ether enjoys a variety of niche applications.

Mesitylene

In organic chemistry, Mesitylene or 1,3,5-trimethylbenzene (C_9H_{12}) is an aromatic hydrocarbon with three methyl substituents attached to the benzene ring. It is prepared from distillation of acetone with sulfuric acid or by the trimerization of propyne in sulfuric acid, which in both cases acts as a catalyst and dehydrating agent. It is commonly used as a solvent in research and industry.

Mutarotation

Mutarotation is the term given to the change in the specific rotation of a cyclic monosaccharide as it reaches an equilibrium between its α and β anomeric forms. It was discovered by Dubrunfaut in 1846, when he noticed that the specific rotation of aqueous sugar solution changes with time. The optical rotation of the solution depends on the optical rotation of each anomer and their ratio in the solution.

Phenyl group

In organic chemistry, the Phenyl group or phenyl ring is the functional group with the formula

C_6H_5-,

where the six carbon atoms are arranged in a cyclic ring structure. This hydrophobic, highly-stable and aromatic hydrocarbon unit can be found in many organic compounds. It can be thought of as being derived from benzene .

Esters

Esters are chemical compounds derived formally from an oxoacid (one containing an oxo group, X=O), and a hydroxyl compound such as an alcohol or phenol. Esters consist of an inorganic acid or organic acid in which at least one -OH (hydroxyl) group is replaced by an -O-alkyl (alkoxy) group. They are analogous to salts, using organic alcohols instead of metallic hydroxides.

Iodide

An Iodide ion is an iodine atom with a −1 charge. Compounds with iodine in formal oxidation state −1 are called Iodide s. This can include ionic compounds such as caesium Iodide or covalent compounds such as phosphorus tri Iodide .

M-xylene

M-Xylene is an aromatic hydrocarbon, based on benzene with two methyl substituents.

It is an isomer of o-xylene and p-xylene. The m stands for meta, meaning the two methyl substituents are at locants 1 and 3 on the aromatic ring.

O-toluic acid	O-Toluic acid, also 2-methylbenzoic acid, is an aromatic carboxylic acid, with formula $(CH_3)C_6H_4(COOH.)$ It is an isomer of p-toluic acid and m-toluic acid. When purified and recrystallized, O-Toluic acid forms needle-shaped crystals.
P-cresol	P-Cresol, also 4-methylphenol, is a phenol, with formula $(CH_3)C_6H_4(OH.)$ It is a positional isomer; the other two are m-cresol and o-cresol. P-Cresol is a major component in pig odor .
Bromide	A Bromide ion is a bromine atom with charge of −1. Compounds with bromine in formal oxidation state −1 are called Bromide s, and each individual chemical in this class can be called a Bromide as well. The class name can include ionic compounds such as caesium Bromide or covalent compounds such as sulfur di Bromide .

Electrophilic aromatic substitution	Electrophilic aromatic substitution or Electrophilic aromatic substitution is an organic reaction in which an atom, usually hydrogen, appended to an aromatic system is replaced by an electrophile. The most important reactions of this type that take place are aromatic nitration, aromatic halogenation, aromatic sulfonation, and acylation and alkylating Friedel-Crafts reactions. Aromatic nitrations to form nitro compounds take place by generating a nitronium ion from nitric acid and sulfuric acid.
Toluene	Toluene phenylmethane, and Toluol, is a clear water-insoluble liquid with the typical smell of paint thinners, redolent of the sweet smell of the related compound benzene. It is an aromatic hydrocarbon that is widely used as an industrial feedstock and as a solvent. Like other solvents, Toluene is also used as an inhalant drug for its intoxicating properties; however this causes severe neurological harm.
Benzene	Benzene, or benzol, is an organic chemical compound and a known carcinogen with the molecular formula C_6H_6. It is sometimes abbreviated Ph-H. Benzene is a colorless and highly flammable liquid with a sweet smell and a relatively high melting point. Because it is a known carcinogen, its use as an additive in gasoline is now limited, but it is an important industrial solvent and precursor in the production of drugs, plastics, synthetic rubber, and dyes.
Alkynes	Alkynes are hydrocarbons that have a triple bond between two carbon atoms, with the formula C_nH_{2n-2}. Alkynes are traditionally known as acetylenes, although the name acetylene also refers specifically to C_2H_2, known formally (but rarely) as ethyne using IUPAC nomenclature. Like other hydrocarbons, Alkynes are generally hydrophobic but tend to be more reactive.
Iodobenzene	Iodobenzene is an organic compound consisting of a benzene ring substitituted with one iodine atom. It is useful as a synthetic intermediate in organic chemistry. Since the C-I bond is weaker than C-Br or C-Cl, it is more reactive than bromobenzene or chlorobenzene.
Nitrobenzene	Nitrobenzene, also known as nitrobenzol or oil of mirbane, is an organic compound with the chemical formula $C_6H_5NO_2$. Nitrobenzene is a water-insoluble oil which exhibits a pale yellow to yellow-brown coloration in liquid form (at room temperature and pressure) with an almond-like odor. When frozen, it appears as a greenish-yellow crystal.
5-nitro-2-propoxyaniline	5-Nitro-2-propoxyaniline is one of the strongest sweet-tasting substances known, about 4,000 times the intensity of sucrose. It is an orange solid that is only slightly soluble in water. It is stable in boiling water and dilute acids.
Nitronium ion	The Nitronium ion, NO_2^+, is a generally unstable cation created by the removal of an electron from the paramagnetic nitrogen dioxide molecule, or the protonation of nitric acid. It is not stable enough to exist in normal conditions, but it is used extensively as an electrophile in the nitration of other substances. The ion is generated in situ for this purpose by mixing sulfuric acid and nitric acid according to the equilibrium: $2\ H_2SO_4 + HNO_3 \rightarrow 2\ HSO_4^- + NO_2^+ + H_3O^+$ The Nitronium ion also exists in the solid form of dinitrogen pentoxide, N_2O_5, which is an ionic solid formed from nitronium and nitrate ions.

Benzenesulfonic acid	Benzenesulfonic acid is an organic compound with the formula $C_6H_5SO_3H$. It is the simplest aromatic sulfonic acid. It is a colourless deliquescent sheet crystal or white wax solid that is soluble in water and ethanol, slightly soluble in benzene and insoluble in carbon disulfide and diethyl ether. It is usually stored in the form of alkali metal salts.
Activating group	In organic chemistry, a functional group is called an Activating group if a benzene molecule to which it is attached more readily participates in electrophilic substitution reactions. Benzene itself will normally undergo substitutions by electrophiles, but additional substituents can alter the reaction rate or products by electronically or sterically affecting the interaction of the two reactants. Functional groups are typically divided into three levels of activating ability.
Nitration	Nitration is a general chemical process for the introduction of a nitro group into a chemical compound. Examples of nitrations are the conversion of glycerin to nitroglycerin and the conversion of toluene to trinitrotoluene. Both of these conversions use nitric acid and sulfuric acid .
Benzocaine	Benzocaine is a local anesthetic commonly used as a topical pain reliever. It is the active ingredient in many over-the-counter anesthetic ointments (e.g. products for oral ulcers of Anbesol by Wyeth, Kank+a by Blistex, Orabase B and Orajel by Del Pharmaceuticals, and Ultracare by Ultradent.) It is also combined with antipyrine to form A/B Otic Drops, (Brand name Auralgan) to relieve earpain and remove cerumen.
Procaine	Procaine is a local anesthetic drug of the amino ester group. It is used primarily to reduce the pain of intramuscular injection of penicillin, and it is also used in dentistry. Owing to the ubiquity of the trade name Novocain, Procaine is sometimes referred to generically as novocaine.
Alkyl	In chemistry, an Alkyl group is a hydrocarbon; typically an Alkyl is a part of a larger molecule. The term is usually used loosely, there is no general formula for an Alkyl group. In structural formulae, an Alkyl group is represented with an R. Usually, Alkyl groups resemble hydrocarbons, but with one less hydrogen atom.
Ethylbenzene	Ethylbenzene is an organic compound with the formula $C_6H_5CH_2CH_3$. This aromatic hydrocarbon is important in the petrochemical industry as an intermediate in the production of styrene, which in turn is used for making polystyrene, a commonly used plastic material. Although often present in small amounts in crude oil, Ethylbenzene is produced in bulk quantities by combining benzene and ethylene in an acid-catalyzed chemical reaction: $C_6H_6 + C_2H_4 \rightarrow C_6H_5CH_2CH_3$ Approximately 24,700,000 tons were produced in 1999.
Inductive effect	The Inductive effect in chemistry is an experimentally observable effect of the transmission of charge through a chain of atoms in a molecule by electrostatic induction. The net polar effect exerted by a substituent is a combination of this Inductive effect and the mesomeric effect. The electron cloud in a σ-bond between two unlike atoms is not uniform and is slightly displaced towards the more electronegative of the two atoms.

Substituent	In organic chemistry and biochemistry, a Substituent is an atom or group of atoms substituted in place of a hydrogen atom on the parent chain of a hydrocarbon. The terms Substituent, side chain, group, branch, or pendant group are used almost interchangeably to describe branches from a parent structure, though certain distinctions are made in the context of polymer chemistry. In polymers, side chains extend from a backbone structure.
Alkene	In organic chemistry, an Alkene olefin, or olefine is an unsaturated chemical compound containing at least one carbon-to-carbon double bond. The simplest acyclic Alkene s, with only one double bond and no other functional groups, form a homologous series of hydrocarbons with the general formula C_nH_{2n}. The simplest Alkene is ethylene (C_2H_4), which has the International Union of Pure and Applied Chemistry (IUPAC) name ethene.
Anisole	Anisole is the organic compound with the formula $CH_3OC_6H_5$. This colorless liquid has a smell reminiscent of anise seed. It is used as a precursor to other organic compounds.
Configuration	The configuration of a molecule is the permanent geometry that results from the spatial arrangement of its bonds. The ability of the same set of atoms to form two or more molecules with different configurations is stereoisomerism. configuration is distinct from chemical conformation, a shape attainable by bond rotations.
Amine	Amine s are organic compounds and functional groups that contain a basic nitrogen atom with a lone pair. Amine s are derivatives of ammonia, wherein one or more hydrogen atoms have been replaced by a substituent such as an alkyl or aryl group. Important Amine s include amino acids, biogenic Amine s, trimethyl Amine , and aniline; see Category: Amine s for a list of Amine s.
Aniline	Aniline, phenylamine or aminobenzene is an organic compound with the formula C_6H_7N. It is the simplest and one of the most important aromatic amines, being used as a precursor to more complex chemicals. Its main application is in the manufacture of polyurethane. Like most volatile amines, it possesses the somewhat unpleasant odour of rotten fish and also has a burning aromatic taste; it is a highly-acrid poison.
Alkanes	Alkanes are chemical compounds that consist only of the elements carbon and hydrogen, wherein these atoms are linked together exclusively by single bonds without any cyclic structure Alkanes belong to a homologous series of organic compounds in which the members differ by a constant relative atomic mass of 14. Each carbon atom must have 4 bonds, and each hydrogen atom must be joined to a carbon atom
Xylene	The term Xylene or xylol refers to a mixture of three aromatic hydrocarbon isomers which is used as a solvent in the printing, rubber, and leather industries. Xylene is a clear, colorless, sweet-smelling liquid that is very flammable. It is usually refined from crude oil in a process called alkylation.
Alkylation	Alkylation is the transfer of an alkyl group from one molecule to another. The alkyl group may be transferred as an alkyl carbocation, a free radical, a carbanion or a carbene (or their equivalents) . Alkylating agents are widely used in chemistry because the alkyl group is probably the most common group encountered in organic molecules.

Carbocation	A Carbocation is an ion with a positively-charged carbon atom. The charged carbon atom in a Carbocation is a 'sextet', i.e. it has only six electrons in its outer valence shell instead of the eight valence electrons that ensures maximum stability . Therefore Carbocation s are often reactive, seeking to fill the octet of valence electrons as well as regain a neutral charge.
Acyl	An Acyl group (IUPAC name: alkanoyl) is a functional group derived by the removal of one or more hydroxyl groups from an oxoacid. In organic chemistry, the Acyl group is usually derived from a carboxylic acid. It therefore has the formula RC(=O)-, where R represents an alkyl group; there is a double bond between the carbon and oxygen atoms, and a single bond between R and the carbon.
Acylation	In chemistry, Acylation is the process of adding an acyl group to a compound. The compound providing the acyl group is called the acylating agent. Because they form a strong electrophile when treated with some metal catalysts, acyl halides are commonly used as acylating agents.
Clemmensen reduction	The Clemmensen reduction is a chemical reaction described as a reduction of ketones (or aldehydes) to alkanes using zinc amalgam and hydrochloric acid. This reaction is named after Erik Christian Clemmensen, a Danish chemist. Several reviews have been published.
Benzaldehyde	Benzaldehyde is a chemical compound consisting of a benzene ring with an aldehyde substituent. It is the simplest representative of the aromatic aldehydes and one of the most industrially used members of this family of compounds. At room temperature it is a colorless liquid with a characteristic and pleasant almond-like odor: Benzaldehyde is an important component of the scent of almonds, hence its typical odor.
Petroleum	Petroleum or crude oil is a naturally occurring, flammable liquid found in rock formations in the Earth consisting of a complex mixture of hydrocarbons of various molecular weights, plus other organic compounds. The term 'Petroleum' was first used in the treatise De Natura Fossilium, published in 1546 by the German mineralogist Georg Bauer, also known as Georgius Agricola. The proportion of hydrocarbons in the mixture is highly variable and ranges from as much as 97% by weight in the lighter oils to as little as 50% in the heavier oils and bitumens.
Nucleophilic substitution	In organic and inorganic chemistry, Nucleophilic substitution is a fundamental class of substitution reaction in which an 'electron rich' nucleophile selectively bonds with or attacks the positive or partially positive charge of an atom attached to a group or atom called the leaving group; the positive or partially positive atom is referred to as an electrophile. The most general form for the reaction may be given as Nuc: + R-LG → R-Nuc + LG:

The electron pair (:) from the nucleophile (Nuc) attacks the substrate (R-LG) forming a new bond, while the leaving group (LG) departs with an electron pair. The principal product in this case is R-Nuc.

Benzyne

The simplest aryne, C_6H_4 (labeled 1 in the diagram to the right), is sometimes called Benzyne. However, this name is open for criticism because it implies a triple bond which would be a special case of triple bonds, so a better name is didehydrobenzene. Benzyne is, like benzene, stabilized by resonance between structures 1 and 2-- so, like the case of benzene, a better representation of the electronic structure would be 3.

Birch reduction

The Birch reduction is the organic reduction of aromatic rings with sodium and an alcohol in liquid ammonia to form 1,4-cyclohexadienes. The reaction was reported by the Australian chemist Arthur John Birch (1915-1995) in 1944. This reaction provides an alternative to catalytic hydrogenation, which usually reduces the aromatic ring all the way to a cyclohexane (after the initial reduction to a cyclohexadiene, catalytic reduction of the remaining (nonaromatic) double bonds is easier than the first reduction.)

Addition reaction

An Addition reaction in organic chemistry, is in its simplest terms an organic reaction where two or more molecules combine to form a larger one.

There are two main types of polar Addition reaction s:

- Electrophilic addition
- Nucleophilic addition

Other non-polar Addition reaction s exists as well:

- Free radical addition

Addition reaction s are limited to chemical compounds that have multiply-bonded atoms:

- Molecules with carbon-carbon double bonds (alkenes) or triple bonds (alkynes)
- Molecules with carbon - hetero double bonds like C=O or C=N

An Addition reaction is the opposite of an elimination reaction. For instance the hydration reaction of an alkene and the dehydration of an alcohol are addition-elimination pairs.

In the related Addition-elimination reaction an Addition reaction is followed by an elimination reaction.

1,3-cyclohexadiene

1,3-Cyclohexadiene is a highly flammable cycloalkene that occurs as a colorless clear liquid. Its refractive index is 1.475 (20 °C, D.)

It can be used as a hydrogen donor in transfer hydrogenation, since its conversion to benzene + hydrogen is in fact exothermic (−24.3 kJ/mol in the gas phase, as indicated by heats of hydrogenation.)

1,4-cyclohexadiene

1,4-Cyclohexadiene is a highly flammable cycloalkene that occurs as a colorless clear liquid.

1,4-Cyclohexadiene and related compounds may be prepared from benzene using lithium or sodium in liquid ammonia, this process being known as a Birch reduction. However 1,4-Cyclohexadiene is easily oxidised to benzene, the driving force being the formation of an aromatic ring.

Potassium permanganate

Potassium permanganate is the inorganic chemical compound $KMnO_4$, a water soluble salt consisting of equal mole amounts of potassium (K^+) and permanganate (MnO_4^-, officially called manganate (VII)) ions. This salt, formerly known as permanganate of potash or Condy's crystals is a strong oxidizing agent. It dissolves in water to give deep purple solutions, evaporation of which gives prismatic purplish-black glistening crystals.

Alcohol

In chemistry, an Alcohol is any organic compound in which a hydroxyl group (-OH) is bound to a carbon atom of an alkyl or substituted alkyl group. The general formula for a simple acyclic Alcohol is $C_nH_{2n+1}OH$. In common terms, the word Alcohol refers to ethanol, the type of Alcohol found in Alcohol ic beverages.

Ethanol is a colorless, volatile liquid with a mild odor which can be obtained by the fermentation of sugars.

Halogenation

Halogenation is a chemical reaction that incorporates a halogen atom into a molecule. More specific descriptions exist that specify the type of halogen: fluorination, chlorination, bromination, and iodination.

In a Markovnikov addition reaction, a halogen like bromine is reacted with an alkene which causes the π-bond to break forming an haloalkane.

Halide

A Halide is a binary compound, of which one part is a halogen atom and the other part is an element or radical that is less electronegative than the halogen, to make a fluoride, chloride, bromide, iodide, or astatide compound. Many salts are Halide s. All Group 1 metals form Halide s with the halogens and they are white solids.

Aspirin

Aspirin is a salicylate drug, often used as an analgesic to relieve minor aches and pains, as an antipyretic to reduce fever, and as an anti-inflammatory medication.

Aspirin also has an antiplatelet effect by inhibiting thromboxane prostaglandins, which under normal circumstances bind platelet molecules together to repair damaged blood vessels. This is why Aspirin is used in long-term, low doses to prevent heart attacks, strokes, and blood clot formation in people at high risk for developing blood clots.

Phenol

In organic chemistry, phenols, sometimes called phenolics, are a class of chemical compounds consisting of a hydroxyl group (-OH) bonded directly to an aromatic hydrocarbon group. The simplest of the class is Phenol Phenol - the simplest of the phenols.

Although similar to alcohols, phenols have unique properties and are not classified as alcohols (since the hydroxyl group is not bonded to a saturated carbon atom.)

Quinone

Quinones are 'compounds having a fully conjugated cyclic dione structure, such as that of benzoquinones, derived from aromatic compounds by conversion of an even number of -CH= groups into -C(=O)- groups with any necessary rearrangement of double bonds (polycyclic and heterocyclic analogues are included.)'

Benzoquinone, sometimes referred to simply as 'Quinone', is one of the two isomers of cyclohexadienedione. These compounds have the molecular formula $C_6H_4O_2$. Orthobenzoquinone is the 1,2-dione, whereas parabenzoquinone is the 1,4-dione.

Aldehyde

An Aldehyde is an organic compound containing a terminal carbonyl group. This functional group, which consists of a carbon atom bonded to a hydrogen atom and double-bonded to an oxygen atom (chemical formula O=CH-), is called the Aldehyde group. The Aldehyde group is also called the formyl or methanoyl group.

Ketone

In organic chemistry, a Ketone is a type of compound that features a carbonyl group bonded to two other carbon atoms, i.e., $R_3CCO\text{-}CR_3$ where R can be a variety of atoms and groups of atoms. With carbonyl carbon bonded to two carbon atoms, Ketone s are distinct from many other functional groups, such as carboxylic acids, aldehydes, esters, amides, and other oxygen-containing compounds. The double-bond of the carbonyl group distinguishes Ketone s from alcohols and ethers.

Alcohol

In chemistry, an Alcohol is any organic compound in which a hydroxyl group (-OH) is bound to a carbon atom of an alkyl or substituted alkyl group. The general formula for a simple acyclic Alcohol is $C_nH_{2n+1}OH$. In common terms, the word Alcohol refers to ethanol, the type of Alcohol found in Alcohol ic beverages.

Ethanol is a colorless, volatile liquid with a mild odor which can be obtained by the fermentation of sugars.

Acetaldehyde

Acetaldehyde is an organic chemical compound with the formula CH_3CHO or MeCHO. It is a flammable liquid. Acetaldehyde occurs naturally in ripe fruit, coffee, and bread, and is produced by plants as part of their normal metabolism. It is popularly known as a chemical that causes hangovers.

Formaldehyde

Formaldehyde (IUPAC name methanal) is a chemical compound with the formula CH_2O. It is the simplest aldehyde. Formaldehyde also exists as the cyclic trimer trioxane and the polymer para Formaldehyde . It exists in water as the hydrate $H_2C(OH)_2$.

Paraformaldehyde

Paraformaldehyde is the condensation product of formaldehyde with a typical chain length of 8 - 100 units. Long chain-length Paraformaldehyde is used as a thermoplastic and is known as polyoxymethylene plastic Paraformaldehyde commonly has a slight odor of formaldehyde due to decomposition.

Trioxane

Trioxane refers to a pair of isomeric organic compounds having the molecular formula $C_3H_6O_3$. Each contains a six membered ring with three carbon atoms and three oxygen atoms.

The two isomers are:

- 1,2,4-Trioxane, an hypothetical compound that occurs as a structural element of some antimalarial agents:
- 1,3,5-Trioxane, a trimer of formaldehyde used as fuel and in plastics manufacture:

.

Aldol

An Aldol or Aldol adduct is a beta-hydroxy ketone or aldehyde, and is the product of Aldol addition (as opposed to Aldol condensation, which produces an α,β-unsaturated carbonyl moiety.)

Typically, 'Aldol' refers to 3-hydroxybutanal.

Aldol condensation

An Aldol condensation is an organic reaction in which an enolate ion reacts with a carbonyl compound to form a β-hydroxyaldehyde or β-hydroxyketone, followed by dehydration to give a conjugated enone.

	Aldol condensation s are important in organic synthesis, providing a good way to form carbon-carbon bonds. The Robinson annulation reaction sequence features an Aldol condensation the Wieland-Miescher ketone product is an important starting material for many organic syntheses.
Metaldehyde	Metaldehyde is a chemical that is commonly used as a pesticide against slugs, snails, and other gastropods. Metaldehyde is a cyclic tetramer of acetaldehyde, with IUPAC name: r-2, c-4, c-6, c-8-tetramethyl-1,3,5,7-tetroxoctane. It is sold under various trade names as a molluscicide, including Antimilice, Ariotox, Cekumeta, Deadline, Halizan, Limatox, Limeol, Meta, Metason, Mifaslug, Namekil, Slug Fest Colloidel 25, and Slugit.
Paraldehyde	Paraldehyde is the cyclic trimer of acetaldehyde molecules. Formally, it is a derivative of 1,3,5-trioxane. The corresponding tetramer is metaldehyde.
Butyraldehyde	Butyraldehyde, also known as butanal, is an organic compound with the formula $CH_3(CH_2)_2CHO$. This compound is the aldehyde derivative of butane. It is a colourless flammable liquid with an acrid smell. It is miscible with most organic solvents.
Carbon	Carbon is the chemical element with symbol C and atomic number 6. As a member of group 14 on the periodic table, it is nonmetallic and tetravalent--making four electrons available to form covalent chemical bonds. There are three naturally occurring isotopes, with ^{12}C and ^{13}C being stable, while ^{14}C is radioactive, decaying with a half-life of about 5730 years.
McLafferty rearrangement	The McLafferty rearrangement is a reaction observed in mass spectrometry. It is sometimes found that a molecule containing a keto-group undergoes β-cleavage, with the gain of the γ-hydrogen atom. This rearrangement may take place by a radical or ionic mechanism.
Ozonolysis	Ozonolysis is the cleavage of an alkene or alkyne with ozone to form organic compounds in which the multiple carbon-carbon bond has been replaced by a double bond to oxygen. The outcome of the reaction depends on the type of multiple bond being oxidized and the workup conditions. Alkenes can be oxidized with ozone to form alcohols, aldehydes or ketones, or carboxylic acids.
Pyridinium	Pyridinium refers to the cationic form of pyridine. This can either be due to protonation of the ring nitrogen or because of addition of a substituent to the ring nitrogen, typically via alkylation. The lone pair of electrons on the nitrogen atom of pyridine is not delocalized, and thus pyridine can be protonated easily.
Pyridinium chlorochromate	Pyridinium chlorochromate is a reddish orange solid reagent used to oxidize primary alcohols to aldehydes and secondary alcohols to ketones. Pyridinium chlorochromate, or Pyridinium chlorochromateC, will not fully oxidize the alcohol to the carboxylic acid as does the Jones reagent. A disadvantage to using Pyridinium chlorochromateC is its toxicity.
Alkene	In organic chemistry, an Alkene olefin, or olefine is an unsaturated chemical compound containing at least one carbon-to-carbon double bond. The simplest acyclic Alkene s, with only one double bond and no other functional groups, form a homologous series of hydrocarbons with the general formula C_nH_{2n}.

The simplest Alkene is ethylene (C_2H_4), which has the International Union of Pure and Applied Chemistry (IUPAC) name ethene.

Alkynes

Alkynes are hydrocarbons that have a triple bond between two carbon atoms, with the formula C_nH_{2n-2}. Alkynes are traditionally known as acetylenes, although the name acetylene also refers specifically to C_2H_2, known formally (but rarely) as ethyne using IUPAC nomenclature. Like other hydrocarbons, Alkynes are generally hydrophobic but tend to be more reactive.

Acylation

In chemistry, Acylation is the process of adding an acyl group to a compound. The compound providing the acyl group is called the acylating agent.

Because they form a strong electrophile when treated with some metal catalysts, acyl halides are commonly used as acylating agents.

Addition reaction

An Addition reaction in organic chemistry, is in its simplest terms an organic reaction where two or more molecules combine to form a larger one.

There are two main types of polar Addition reaction s:

- Electrophilic addition
- Nucleophilic addition

Other non-polar Addition reaction s exists as well:

- Free radical addition

Addition reaction s are limited to chemical compounds that have multiply-bonded atoms:

- Molecules with carbon-carbon double bonds (alkenes) or triple bonds (alkynes)
- Molecules with carbon - hetero double bonds like C=O or C=N

An Addition reaction is the opposite of an elimination reaction. For instance the hydration reaction of an alkene and the dehydration of an alcohol are addition-elimination pairs.

In the related Addition-elimination reaction an Addition reaction is followed by an elimination reaction.

Petroleum

Petroleum or crude oil is a naturally occurring, flammable liquid found in rock formations in the Earth consisting of a complex mixture of hydrocarbons of various molecular weights, plus other organic compounds.

The term 'Petroleum' was first used in the treatise De Natura Fossilium, published in 1546 by the German mineralogist Georg Bauer, also known as Georgius Agricola.

	The proportion of hydrocarbons in the mixture is highly variable and ranges from as much as 97% by weight in the lighter oils to as little as 50% in the heavier oils and bitumens.
Carboxylic acids	Carboxylic acids are organic acids characterized by the presence of a carboxyl group, which has the formula -C(=O)OH, usually written -COOH or -CO_2H. Carboxylic acids are Brønsted-Lowry acids -- they are proton donors. Salts and anions of Carboxylic acids are called carboxylates. The simplest series of Carboxylic acids are the alkanoic acids, R-COOH, where R is a hydrogen or an alkyl group.
Organolithium reagent	An Organolithium reagent is an organometallic compound with a direct bond between a carbon and a lithium atom. As the electropositive nature of lithium puts most of the charge density of the bond on the carbon atom, effectively creating a carbanion, organolithium compounds are extremely powerful bases and nucleophiles. Organolithium reagents are industrially prepared by the reaction of an halocarbon with lithium metal, i.e. R-X + 2 Li $\rightarrow$ R-Li + LiX. A side reaction of this synthesis, especially with alkyl iodides, is the Wurtz reaction, in which an R-Li species reacts with an R-X species forming an R-R coupled product.
Benzophenone	Benzophenone is the organic compound with the formula $(C_6H_5)_2CO$, generally abbreviated Ph_2CO. Benzophenone is a widely used building block in organic chemistry, being the parent diarylketone. Benzophenone can be used as a photo initiator in UV-curing applications such as inks, imaging, and clear coatings in the printing industry. Benzophenone prevents ultraviolet (UV) light from damaging scents and colors in products such as perfumes and soaps.
Cycloalkanes	Cycloalkanes are types of alkanes which have one or more rings of carbon atoms in the chemical structure of their molecules. Alkanes are types of organic hydrocarbon compounds which have only single chemical bonds in their chemical structure. Cycloalkanes consist of only carbon and hydrogen atoms and are saturated because there are no multiple C-C bonds to hydrogenate
Nitrile	A Nitrile is any organic compound which has a -C≡N functional group. The -C≡N functional group is called a Nitrile group. In the -CN group, the carbon atom and the nitrogen atom are triple bonded together.
Gilman reagent	A Gilman reagent is a lithium and copper (diorganocopper) reagent compound, R_2CuLi, where R is an organic radical. These are useful because they react with chlorides, bromides, and iodides to replace the halide group with an R group. This is extremely useful in creating larger molecules from smaller ones.
Electrophilic addition	In organic chemistry, an Electrophilic addition reaction is an addition reaction where, in a chemical compound, a pi bond is removed by the creation of two new covalent bonds. The substrate of an Electrophilic addition reaction must have a double bond or triple bond . The driving force for this reaction is the formation of an electrophile X^+ that forms a covalent bond with an electron-rich unsaturated C=C bond.

Nucleophilic addition	In organic chemistry, a Nucleophilic addition reaction is an addition reaction where in a chemical compound a π bond is removed by the creation of two new covalent bonds by the addition of a nucleophile . Addition reactions are limited to chemical compounds that have multiple-bonded atoms: • molecules with carbon - hetero multiple bonds like carbonyls, imines or nitriles • molecules with carbon - carbon double bonds or triple bonds Addition reactions of a nucleophile to carbon - hetero double bonds such as C=O or CN triple bond show a wide variety. These bonds are polar (have a large difference in electronegativity between the two atoms) consequently carbon carries a partial positive charge. This makes this atom the primary target for the nucleophile.
Electronic effect	An Electronic effect influences the structure, reactivity, or properties of molecule but is neither a traditional bond nor a steric effect. In organic chemistry, the term stereo Electronic effect is also used to emphasize the relation between the electronic structure and the geometry (stereochemistry) of a molecule. Induction is the redistribution of electron density through a traditional sigma bonded structure according to the electronegativity of the atoms involved.
Cyanohydrin	A Cyanohydrin is a functional group found in organic compounds. Cyanohydrin s have the formula $R_2C(OH)CN$, where R is H, alkyl, or aryl. Cyanohydrin s are industrially important precursors to carboxylic acids and some amino acids.
Wittig reaction	The Wittig reaction is a chemical reaction of an aldehyde or ketone with a triphenyl phosphonium ylide (often called a Wittig reagent) to give an alkene and triphenylphosphine oxide. The Wittig reaction was discovered in 1954 by Georg Wittig, for which he was awarded the Nobel Prize in Chemistry in 1979. It is widely used in organic synthesis for the preparation of alkenes.
Betaine	A Betaine in chemistry is any neutral chemical compound with a positively charged cationic functional group such as an ammonium ion or phosphonium ion (generally: onium ions) which bears no hydrogen atom and with a negatively charged functional group such as a carboxylate group which may not be adjacent to the cationic site. A Betaine thus may be a specific type of zwitterion. Historically the term was reserved for trimethylglycine only.

Ketone

In organic chemistry, a Ketone is a type of compound that features a carbonyl group bonded to two other carbon atoms, i.e., $R_3CCO\text{-}CR_3$ where R can be a variety of atoms and groups of atoms. With carbonyl carbon bonded to two carbon atoms, Ketone s are distinct from many other functional groups, such as carboxylic acids, aldehydes, esters, amides, and other oxygen-containing compounds. The double-bond of the carbonyl group distinguishes Ketone s from alcohols and ethers.

Aldehyde

An Aldehyde is an organic compound containing a terminal carbonyl group. This functional group, which consists of a carbon atom bonded to a hydrogen atom and double-bonded to an oxygen atom (chemical formula O=CH-), is called the Aldehyde group. The Aldehyde group is also called the formyl or methanoyl group.

Chloral

Chloral, also known as trichloroacetaldehyde, is the organic compound with the formula Cl_3CCHO. This aldehyde is a colourless oily liquid that is soluble in a wide range of solvents. It reacts with water to form Chloral hydrate, a once widely used sedative and hypnotic substance.

Chloral can be produced by chlorination of ethanol, as reported in 1832 by Justus von Liebig.

Cyanohydrin

A Cyanohydrin is a functional group found in organic compounds. Cyanohydrin s have the formula $R_2C(OH)CN$, where R is H, alkyl, or aryl. Cyanohydrin s are industrially important precursors to carboxylic acids and some amino acids.

Hydrogen

Hydrogen is the chemical element with atomic number 1. It is represented by the symbol H. At standard temperature and pressure, Hydrogen is a colorless, odorless, nonmetallic, tasteless, highly flammable diatomic gas with the molecular formula H_2. With an atomic weight of 1.007 94 u, Hydrogen is the lightest element.

Nitrile

A Nitrile is any organic compound which has a -C≡N functional group. The -C≡N functional group is called a Nitrile group. In the -CN group, the carbon atom and the nitrogen atom are triple bonded together.

Imine

An Imine is a functional group or chemical compound containing a carbon-nitrogen double bond . Due to their diverse reactivity, Imine s are common substrates in a wide variety of transformations. An Imine can be synthesised by the nucleophilic addition of an amine to a ketone or aldehyde giving a hemiaminal -C(OH)(NHR)- followed by an elimination of water to yield the Imine

Schiff base

A Schiff base is a functional group that contains a carbon-nitrogen double bond with the nitrogen atom connected to an aryl or alkyl group--but not hydrogen. Schiff bases are of the general formula $R_1R_2C\text{=}N\text{-}R_3$, where R_3 is an aryl or alkyl group that makes the Schiff base a stable imine. A Schiff base derived from an aniline, where R_3 is a phenyl or substituted phenyl, can be called an anil.

Hydrazine

Hydrazine is an inorganic chemical compound with the formula N_2H_4. It is a colourless liquid with an ammonia-like odor and is derived from the same industrial chemistry processes that manufacture ammonia. However, Hydrazine has physical properties that are more similar to those of water.

Hydrazone

A Hydrazone is a class of organic compounds with the structure $R_2C\text{=}NNR_2$. They are related to ketones and aldehydes by the replacement of the oxygen with the NNR_2 functional group. They are formed usually by the action of hydrazine on ketones or aldehydes.

Oxime	An Oxime is one in a class of chemical compounds with the general formula R_1R_2CNOH, where R_1 is an organic side chain and R_2 is either hydrogen, forming an aldoxime, or another organic group, forming a ketoxime. O-substituted oximes form a closely related family of compounds. Amidoximes are oximes of hemiaminals with general structure RC(=NOH)(NRR'.)
Semicarbazone	In organic chemistry, a Semicarbazone is a derivative of an aldehyde or ketone formed by a condensation reaction between a ketone or aldehyde and semicarbazide. For ketones: $H_2NNHC(=O)NH_2 + RC(=O)R \rightarrow R_2C=NNHC(=O)NH_2$ For aldehydes: $H_2NNHC(=O)NH_2 + RCHO \rightarrow RCH=NNHC(=O)NH_2$ For example, the Semicarbazone of acetone would have the structure $(CH_3)_2C=NNHC(=O)NH_2$. A thiosemicarbazone is an analog of a Semicarbazone which contains a sulfur atom in place of the oxygen atom.
Acetaldehyde	Acetaldehyde is an organic chemical compound with the formula CH_3CHO or MeCHO. It is a flammable liquid. Acetaldehyde occurs naturally in ripe fruit, coffee, and bread, and is produced by plants as part of their normal metabolism. It is popularly known as a chemical that causes hangovers.
Aldol	An Aldol or Aldol adduct is a beta-hydroxy ketone or aldehyde, and is the product of Aldol addition (as opposed to Aldol condensation, which produces an α,β-unsaturated carbonyl moiety.) Typically, 'Aldol' refers to 3-hydroxybutanal.
Aldol condensation	An Aldol condensation is an organic reaction in which an enolate ion reacts with a carbonyl compound to form a β-hydroxyaldehyde or β-hydroxyketone, followed by dehydration to give a conjugated enone. Aldol condensation s are important in organic synthesis, providing a good way to form carbon-carbon bonds. The Robinson annulation reaction sequence features an Aldol condensation the Wieland-Miescher ketone product is an important starting material for many organic syntheses.
Ethylene	Ethylene is the chemical compound with the formula C_2H_4. It is the simplest alkene. Because it contains a carbon-carbon double bond, Ethylene is called an unsaturated hydrocarbon or an olefin.
Acetal	An Acetal is a molecule with two single bonded oxygens attached to the same carbon atom. Traditional usages distinguish ketal from Acetal (whereas the ketal has two carbon-bonded R groups, the Acetal has one carbon-bonded R group as H-.) Current accepted terminology classifies ketals as a subset of Acetal s.

Carbohydrates	Carbohydrates or saccharides are the most abundant of the four major classes of biomolecules. They fill numerous roles in living things, such as the storage and transport of energy (eg: starch, glycogen) and structural components (eg: cellulose in plants and chitin.) Additionally, Carbohydrates and their derivatives play major roles in the working process of the immune system, fertilization, pathogenesis, blood clotting, and development.
Hemiacetals	Hemiacetals and hemiketals are compounds of the general formula $R_1R'_1C(OH)OR_2$, where R_2 is not hydrogen. These types of compounds are formed from carbonyl compounds and alcohols, i.e. Hemiacetals from aldehydes, hemiketals from ketones. Thus, for Hemiacetals, either R_1 or R'_1 must be a hydrogen, while for hemiketals, neither will be.
Protecting group	A Protecting group or protective group is introduced into a molecule by chemical modification of a functional group in order to obtain chemoselectivity in a subsequent chemical reaction. It plays an important role in multistep organic synthesis. Acetal protection of a ketone during reduction of an ester In many preparations of delicate organic compounds, some specific parts of their molecules cannot survive the required reagents or chemical environments.
Alcohol	In chemistry, an Alcohol is any organic compound in which a hydroxyl group (-OH) is bound to a carbon atom of an alkyl or substituted alkyl group. The general formula for a simple acyclic Alcohol is $C_nH_{2n+1}OH$. In common terms, the word Alcohol refers to ethanol, the type of Alcohol found in Alcohol ic beverages. Ethanol is a colorless, volatile liquid with a mild odor which can be obtained by the fermentation of sugars.
Cracking	In petroleum geology and chemistry, Cracking is the process whereby complex organic molecules such as kerogens or heavy hydrocarbons are broken down into simpler molecules (e.g. light hydrocarbons) by the breaking of carbon-carbon bonds in the precursors. The rate of Cracking and the end products are strongly dependent on the temperature and presence of any catalysts. Cracking, also referred to as pyrolysis, is the breakdown of a large alkane into smaller, more useful alkanes and an alkene.
Hydrogenation	Hydrogenation is the chemical reaction that results from the addition of hydrogen (H_2.) The process is usually employed to reduce or saturate organic compounds. The process typically constitutes the addition of pairs of hydrogen atoms to a molecule.
Clemmensen reduction	The Clemmensen reduction is a chemical reaction described as a reduction of ketones (or aldehydes) to alkanes using zinc amalgam and hydrochloric acid. This reaction is named after Erik Christian Clemmensen, a Danish chemist. Several reviews have been published.

Alkaloids
Alkaloids are naturally occurring chemical compounds containing basic nitrogen atoms. The name derives from the word alkaline and was used to describe any nitrogen-containing base. Alkaloids are produced by a large variety of organisms, including bacteria, fungi, plants, and animals and are part of the group of natural products (also called secondary metabolites.)

Amine
Amine s are organic compounds and functional groups that contain a basic nitrogen atom with a lone pair. Amine s are derivatives of ammonia, wherein one or more hydrogen atoms have been replaced by a substituent such as an alkyl or aryl group. Important Amine s include amino acids, biogenic Amine s, trimethyl Amine , and aniline; see Category: Amine s for a list of Amine s.

Alcohol
In chemistry, an Alcohol is any organic compound in which a hydroxyl group (-OH) is bound to a carbon atom of an alkyl or substituted alkyl group. The general formula for a simple acyclic Alcohol is $C_nH_{2n+1}OH$. In common terms, the word Alcohol refers to ethanol, the type of Alcohol found in Alcohol ic beverages.

Ethanol is a colorless, volatile liquid with a mild odor which can be obtained by the fermentation of sugars.

Aniline
Aniline, phenylamine or aminobenzene is an organic compound with the formula C_6H_7N. It is the simplest and one of the most important aromatic amines, being used as a precursor to more complex chemicals. Its main application is in the manufacture of polyurethane. Like most volatile amines, it possesses the somewhat unpleasant odour of rotten fish and also has a burning aromatic taste; it is a highly-acrid poison.

Mitomycin
The mitomycins are a family of aziridine-containing natural products isolated from Streptomyces lavendulae. One of these compounds, Mitomycin C, finds use as a chemotherapeutic agent by virtue of its antitumour antibiotic activity. It is given intravenously to treat upper gastro-intestinal (e.g. esophageal carcinoma) and breast cancers, as well as by bladder instillation for superficial bladder tumours.

Nitrogen inversion
In chemistry, a nitrogen compound like ammonia in a trigonal pyramid geometry undergoes rapid Nitrogen inversion whereby the molecule turns inside out. This interconversion is a room temperature process because the energy barrier (24.2 kJ/mol) is relatively small. Contrast this to phosphine which does not show inversion at room temperature (energy barrier: 132 kJ/mol) .

Diamine
A Diamine is a type of polyamine with exactly two amino groups.

Examples include:

- The simplest example, hydrazine
- Diamine s with linear carbon chain
    - ethylene Diamine (2 carbon)
    - putrescine (4 carbon)
    - cadaverine (5 carbon)
    - hexamethylene Diamine (6 carbon)
- Other
    - ethambutol
    - phenylene Diamine

.

Cocaine	The first synthesis and elucidation of the cocaine molecule was by Richard Willstätter in 1898. Willstätter's synthesis derived cocaine from tropinone. Since then, Robert Robinson and Edward Leete have made significant contributions to the mechanism of the synthesis.
Ephedrine	Ephedrine is a sympathomimetic amine commonly used as a stimulant, appetite suppressant, concentration aid, decongestant, and to treat hypotension associated with anaesthesia. Ephedrine is similar in structure to the synthetic derivatives amphetamine and methamphetamine. Chemically, it is an alkaloid derived from various plants in the genus Ephedra (family Ephedraceae.)
Butyl	In organic chemistry, Butyl is a four-carbon alkyl substituent with chemical formula $-C_4H_9$. It is derived from either of the two isomers of the alkane called butane. Each of the two isomers of butane give rise to two isomers of the Butyl substituent.
Iminium	An Iminium salt or cation in organic chemistry has the general structure $[R^1R^2C{=}NR^3R^4]^+$ and is as such a protonated or substituted imine . It is an intermediate in many organic reactions such as the Beckmann rearrangement, Vilsmeier-Haack reaction, Stephen reaction or the Duff reaction. The use of the alternative names imonium compounds and immonium compounds is discouraged.
Propyl	In organic chemistry, Propyl is a three-carbon alkyl substituent with chemical formula $-C_3H_7$. It is the substituent form of the alkane propane. For example: Propyl ethanoate, also called Propyl acetate. This is Propyl ethanoate, an ester.
Electrophilic aromatic substitution	Electrophilic aromatic substitution or Electrophilic aromatic substitution is an organic reaction in which an atom, usually hydrogen, appended to an aromatic system is replaced by an electrophile. The most important reactions of this type that take place are aromatic nitration, aromatic halogenation, aromatic sulfonation, and acylation and alkylating Friedel-Crafts reactions. Aromatic nitrations to form nitro compounds take place by generating a nitronium ion from nitric acid and sulfuric acid.
Ketone	In organic chemistry, a Ketone is a type of compound that features a carbonyl group bonded to two other carbon atoms, i.e., $R_3CCO\text{-}CR_3$ where R can be a variety of atoms and groups of atoms. With carbonyl carbon bonded to two carbon atoms, Ketone s are distinct from many other functional groups, such as carboxylic acids, aldehydes, esters, amides, and other oxygen-containing compounds. The double-bond of the carbonyl group distinguishes Ketone s from alcohols and ethers.
Aldehyde	An Aldehyde is an organic compound containing a terminal carbonyl group. This functional group, which consists of a carbon atom bonded to a hydrogen atom and double-bonded to an oxygen atom (chemical formula O=CH-), is called the Aldehyde group. The Aldehyde group is also called the formyl or methanoyl group.

Pyridine	Pyridine is a simple aromatic heterocyclic organic compound with the chemical formula C_5H_5N used as a precursor to agrochemicals and pharmaceuticals, and is also an important solvent and reagent. It is structurally related to benzene, wherein one CH group in the aromatic six-membered ring is replaced by a nitrogen atom. It exists as a colorless liquid with a distinctive, unpleasant fish-like odor.
Nucleophilic substitution	In organic and inorganic chemistry, Nucleophilic substitution is a fundamental class of substitution reaction in which an 'electron rich' nucleophile selectively bonds with or attacks the positive or partially positive charge of an atom attached to a group or atom called the leaving group; the positive or partially positive atom is referred to as an electrophile. The most general form for the reaction may be given as Nuc: + R-LG → R-Nuc + LG: The electron pair (:) from the nucleophile (Nuc) attacks the substrate (R-LG) forming a new bond, while the leaving group (LG) departs with an electron pair. The principal product in this case is R-Nuc.
Alkylation	Alkylation is the transfer of an alkyl group from one molecule to another. The alkyl group may be transferred as an alkyl carbocation, a free radical, a carbanion or a carbene (or their equivalents) . Alkylating agents are widely used in chemistry because the alkyl group is probably the most common group encountered in organic molecules.
Acylation	In chemistry, Acylation is the process of adding an acyl group to a compound. The compound providing the acyl group is called the acylating agent. Because they form a strong electrophile when treated with some metal catalysts, acyl halides are commonly used as acylating agents.
Acyl	An Acyl group (IUPAC name: alkanoyl) is a functional group derived by the removal of one or more hydroxyl groups from an oxoacid. In organic chemistry, the Acyl group is usually derived from a carboxylic acid. It therefore has the formula RC(=O)-, where R represents an alkyl group; there is a double bond between the carbon and oxygen atoms, and a single bond between R and the carbon.
Sulfonamide	In chemistry, the Sulfonamide functional group (also spelt sulphonamide) is $-S(=O)_2-NH_2$, a sulfonyl group connected to an amine group. A Sulfonamide is a compound that contains this group. The general formula is RSO_2NH_2, where R is some organic group.
Hofmann elimination	Hofmann elimination is a process where an amine is reacted to create a tertiary amine and an alkene by treatment with excess methyl iodide followed by treatment with silver oxide, water, and heat . After the first step, a quaternary ammonium iodide salt is created. After replacement of iodine by an hydroxyl anion, an elimination reaction takes place to the alkene.

Leaving group	A Leaving group in chemistry is an ion or substituent with the ability to detach itself from a molecule. The remaining molecule or fragment remaining is known as the residual or main part. The term Leaving group is dependent on the context of the statement.
Methylation	Methylation in the chemical sciences denotes the attachment or substitution of a methyl group on various substrates. This term is commonly used in chemistry, biochemistry, soil science and the biological sciences. In biochemistry, Methylation more specifically refers to the replacement of a hydrogen atom with the methyl group.
Amine oxide	An Amine oxide also known as amine-N-oxide and N-oxide, is a chemical compound that contains the functional group R_3N^+-O^- (sometimes written as R_3N=O or $R_3N \rightarrow O$.) In the strict sense the term Amine oxide applies only to oxides of tertiary amines including nitrogen-containing aromatic compounds like pyridine, but is sometimes also used for the analogous derivatives of primary and secondary amines. Examples of Amine oxide s include pyridine N-oxide, a water-soluble crystalline solid with melting point 62-67°C, and N-methylmorpholine N-oxide, which is an oxidant.
Gabriel synthesis	The Gabriel synthesis is a chemical reaction that transforms primary alkyl halides into primary amines using potassium phthalimide. The sodium or potassium salt of the product reacts with a primary alkyl halide to form an alkyl phthalic imide. The reaction fails with secondary alkyl halides.
Norepinephrine	Noradrenaline (BAN) or Norepinephrine is a catecholamine with dual roles as a hormone and a neurotransmitter. As a stress hormone, Norepinephrine affects parts of the brain where attention and responding actions are controlled. Along with epinephrine, Norepinephrine also underlies the fight-or-flight response, directly increasing heart rate, triggering the release of glucose from energy stores, and increasing blood flow to skeletal muscle.
Serotonin	Serotonin is a monoamine neurotransmitter. It is found extensively in the gastrointestinal tract of animals, and about 80 to 90 percent of the human body's total Serotonin is located in the enterochromaffin cells in the gut, where it is used to regulate intestinal movements. The remainder is synthesized in serotonergic neurons in the central nervous system (CNS) where it has various functions, including control of appetite, mood and anger.
Sandmeyer reaction	The Sandmeyer reaction is a chemical reaction used to synthesize aryl halides from aryl diazonium salts. It is named after the Swiss chemist Traugott Sandmeyer. An aromatic (or heterocyclic) amine quickly reacts with a nitrite to form an aryl diazonium salt, which decomposes in the presence of copper(I) salts, such as copper(I) chloride, to form the desired aryl halide.
Azo compounds	Azo compounds are compounds bearing the functional group R-N=N-R', in which R and R' can be either aryl or alkyl. The N=N group is called an azo group, although the parent compound, HNNH, is called diimide. The more stable derivatives contain two aryl groups.

Diazo

Diazo refers to a type of organic compound that has two linked nitrogen (azo) compounds. The general formula is $R_2C=N_2$. The simplest example of a Diazo compound is Diazo methane.

Strain

In chemistry a molecule experiences Strain when in a chemical conformation there exist unfavorable bond angles or bond distances. Strain energy is released when the molecule can relax to a conformation with less Strain or when the molecule interacts in a suitable chemical reaction.

Several types of Strain exist:

- Angle Strain
- Torsional Strain
- van der Waals Strain

.

Amination

Amination is the process by which an amine group is introduced into an organic molecule. This can occur in a number of ways including reaction with ammonia or another amine such as an alkylation, reductive Amination and the Mannich reaction. For example, -COOH --> $-CONH_2$.

Reductive amination

Reductive amination is a chemical reaction which involves the conversion of a carbonyl group to an amine, via an intermediate imine. The carbonyl group is most commonly a ketone or an aldehyde.

In this organic reaction, the amine first reacts with the carbonyl group to form a hemiaminal species, which subsequently loses one molecule of water in a reversible manner by alkylimino-de-oxo-bisubstitution, to form the imine.

Azide

Azide is the anion with the formula N_3^-. It is the conjugate base of hydrazoic acid. N_3^- is a linear anion that is isoelectronic with CO_2 and N_2O. Per valence bond theory, Azide can be described by several resonance structures, an important one being $N^-=N^+=N^-$.

Nitrile

A Nitrile is any organic compound which has a -C≡N functional group. The -C≡N functional group is called a Nitrile group. In the -CN group, the carbon atom and the nitrogen atom are triple bonded together.

Nitro compounds

Nitro compounds are organic compounds that contain one or more nitro functional groups ($-NO_2$.) They are often highly explosive, especially when the compound contains more than one nitro group. The presence of impurities or improper handling can trigger a violent exothermic decomposition.

Hofmann rearrangement

The Hofmann rearrangement is the organic reaction of a primary amide to a primary amine with one fewer carbon atom.

The reaction of bromine with sodium hydroxide forms sodium hypobromite in situ, which transforms the primary amide into an intermediate isocyanate. The intermediate isocyanate is hydrolyzed to a primary amine giving off carbon dioxide.

Carboxylic acids

Carboxylic acids are organic acids characterized by the presence of a carboxyl group, which has the formula -C(=O)OH, usually written -COOH or $-CO_2H$. Carboxylic acids are Brønsted-Lowry acids -- they are proton donors. Salts and anions of Carboxylic acids are called carboxylates.

The simplest series of Carboxylic acids are the alkanoic acids, R-COOH, where R is a hydrogen or an alkyl group.

Fatty acid

In chemistry, especially biochemistry, a Fatty acid is a carboxylic acid often with a long unbranched aliphatic tail (chain), which is either saturated or unsaturated. Carboxylic acids as short as butyric acid (4 carbon atoms) are considered to be Fatty acid s, whereas Fatty acid s derived from natural fats and oils may be assumed to have at least eight carbon atoms, caprylic acid (octanoic acid), for example. The most abundant natural Fatty acid s have an even number of carbon atoms because their biosynthesis involves acetyl-CoA, a coenzyme carrying a two-carbon-atom group

Fatty acids

Fatty acids are an important source of energy for many organisms. Excess glucose can be stored efficiently as fat. Triglycerides yield more than twice as much energy for the same mass as do carbohydrates or proteins.

Benzoic acid

Benzoic acid, $C_7H_6O_2$ (or C_6H_5COOH), is a colorless crystalline solid and the simplest aromatic carboxylic acid. The name derived from gum benzoin, which was for a long time the only source for Benzoic acid. This weak acid and its salts are used as a food preservative.

Cis-trans isomerism

In organic chemistry, Cis-trans isomerism or geometric isomerism or configuration isomerism or E-Z isomerism is a form of stereoisomerism describing the orientation of functional groups within a molecule. In general, such isomers contain double bonds, which cannot rotate, but they can also arise from ring structures, wherein the rotation of bonds is greatly restricted.

The term 'geometric isomerism' is considered an obsolete synonym of 'Cis-trans isomerism' by IUPAC. It is sometimes used as a synonym for general stereoisomerism (e.g., optical isomerism being called geometric isomerism); the correct term for non-optical stereoisomerism is diastereomerism.

Phthalic acid

Phthalic acid is an aromatic dicarboxylic acid, with formula $C_6H_4(COOH)_2$. It is an isomer of isophthalic acid and terephthalic acid.

Phthalic acid is used mainly in the form of the anhydride to produce other chemicals such as dyes, perfumes, saccharin, phthalates and many others.

Aspirin

Aspirin is a salicylate drug, often used as an analgesic to relieve minor aches and pains, as an antipyretic to reduce fever, and as an anti-inflammatory medication.

Aspirin also has an antiplatelet effect by inhibiting thromboxane prostaglandins, which under normal circumstances bind platelet molecules together to repair damaged blood vessels. This is why Aspirin is used in long-term, low doses to prevent heart attacks, strokes, and blood clot formation in people at high risk for developing blood clots.

Saccharic acid

Saccharic acid, also called glucaric acid, is a chemical compound with the formula $C_6H_{10}O_8$. It is derived by oxidizing a sugar such as glucose with nitric acid.

Source

Source Holdings Ltd. is a specialist European-based provider of Exchange Traded Products (ETPs), including Exchange Traded Funds (ETFs) and Exchange Traded Commodities (ETCs.) The company was founded in 2008 by Bank of America Merrill Lynch, Goldman Sachs and Morgan Stanley and is headquartered inside the City of London, on 88 Wood Street.

Source launched its first product offering on the 20th April 2009, although the 3 founders are reported to have started working on the project in early 2008 .

Carboxylation

Carboxylation in chemistry is a chemical reaction in which a carboxylic acid group is introduced in a substrate. The opposite reaction is de Carboxylation .

In organic chemistry many different protocols exist for Carboxylation

Reagent

A Reagent or reactant is a substance or compound consumed during a chemical reaction. Solvents and catalysts, although they are involved in the reaction, are usually not referred to as reactants.

Although the terms reactant and Reagent are often used interchangeably, a Reagent is more specifically 'a test substance that is added to a system in order to bring about a reaction or to see whether a reaction occurs'.

Nucleophilic acyl Substitution

Nucleophilic acyl substitution describes the substitution reaction involving nucleophiles and acyl compounds. Acyl compounds are carboxylic acid derivatives including esters, amides and acid halides. Nucleophiles include anionic reagents such as alkoxide compounds and enolates or species of high basicity, such as amines .

Esterification

Esterification is the general name for a chemical reaction in which two reactants (typically an alcohol and an acid) form an ester as the reaction product. Esters are common in organic chemistry and biological materials, and often have a characteristic pleasant, fruity odor. This leads to their extensive use in the fragrance and flavor industry.

Diazomethane

Diazomethane is the chemical compound CH_2N_2. It is one of the more common diazo compounds. In the pure form at room temperature, it is a famously explosive yellow gas, but it is almost universally used as a solution in diethyl ether.

Amide

In chemistry, an Amide is one of three kinds of compounds:

- (sometimes called acid Amide the organic functional group characterized by a carbonyl group (C=O) linked to a nitrogen atom (N), or a compound that contains this functional group (pictured to the right); or
- a kind of anion or
- any organic compound derived by the replacement of a hydroxyl group by an amino group.

Amide s are the most stable of all the carbonyl functional groups.

Many chemists make a pronunciation distinction between the two, saying /É™ËˆmiË□d/ for the carbonyl-nitrogen compound and /ËˆæmaÉªd/ for the anion. Others substitute one of these with , while still others pronounce both , making them homonyms.

In the first sense referred to above, an Amide is an amine where one of the nitrogen substituents is an acyl group; it is generally represented by the formula: $R_1(CO)NR_2R_3$, where either or both R_2 and R_3 may be hydrogen.

Carboxylic acids	Carboxylic acids are organic acids characterized by the presence of a carboxyl group, which has the formula -C(=O)OH, usually written -COOH or -CO_2H. Carboxylic acids are Brønsted-Lowry acids -- they are proton donors. Salts and anions of Carboxylic acids are called carboxylates. The simplest series of Carboxylic acids are the alkanoic acids, R-COOH, where R is a hydrogen or an alkyl group.
Geranyl acetate	Geranyl acetate is a natural organic compound that is classified as a monoterpene. It is a colorless liquid with a pleasant floral or fruity rose aroma. Its condensed liquid has a sightly yellow color.
Isoamyl acetate	Isoamyl acetate is an organic compound that is the ester formed from isoamyl alcohol and acetic acid. It is a clear colorless liquid that is only slightly soluble in water, but very soluble in most organic solvents. Isoamyl acetate has a strong odor which is also described as similar to both banana and pear.
Claisen condensation	The Claisen condensation is a carbon-carbon bond forming reaction that occurs between two esters or one ester and another carbonyl compound in the presence of a strong base, resulting in a β-keto ester or a β-diketone. It is named after Rainer Ludwig Claisen, who first published his work on the reaction in 1881 . At least one of the reagents must be enolizable (have an α-proton and be able to undergo deprotonation to form the enolate anion.)
Lactone	In chemistry, a Lactone is a cyclic ester which can be seen as the condensation product of an alcohol group -OH and a carboxylic acid group -COOH in the same molecule. It is characterized by a closed ring consisting of two or more carbon atoms and a single oxygen atom, with a ketone group =O in one of the carbons adjacent to the latter. Lactone nomenclature: α-aceto Lactone , β-propio Lactone , γ-butyro Lactone , and δ-valero Lactone Lactone s are usually named according to the precursor acid molecule (aceto = 2 carbons, propio = 3, butyro = 4, valero = 6, capro = 7, etc.), with a Lactone suffix and a Greek letter prefix that specifies the number of carbons in the heterocyle -- that is, the distance between the relevant -OH and the -COOH groups along said backbone.
Lidocaine	Lidocaine or lignocaine () is a common local anesthetic and antiarrhythmic drug. Lidocaine is used topically to relieve itching, burning and pain from skin inflammations, injected as a dental anesthetic, and in minor surgery. Lidocaine, the first amino amide-type local anesthetic, was first synthesized under the name xylocaine by Swedish chemist Nils Löfgren in 1943.
Amide	In chemistry, an Amide is one of three kinds of compounds: • (sometimes called acid Amide the organic functional group characterized by a carbonyl group (C=O) linked to a nitrogen atom (N), or a compound that contains this functional group (pictured to the right); or • a kind of anion or • any organic compound derived by the replacement of a hydroxyl group by an amino group. Amide s are the most stable of all the carbonyl functional groups.

Many chemists make a pronunciation distinction between the two, saying /É™Ë ̂miË□d/ for the carbonyl-nitrogen compound and /Ë ̂æmaÉªd/ for the anion. Others substitute one of these with , while still others pronounce both , making them homonyms.

In the first sense referred to above, an Amide is an amine where one of the nitrogen substituents is an acyl group; it is generally represented by the formula: $R_1(CO)NR_2R_3$, where either or both R_2 and R_3 may be hydrogen.

Lactam

A Lactam is a cyclic amide. Prefixes indicate how many carbon atoms (apart from the carbonyl moiety) are present in the ring: β-Lactam (2 carbon atoms outside the carbonyl, 4 ring atoms in total), γ-Lactam (3 and 5), δ-Lactam (4 and 6.) Beta β, gamma γ and delta δ are the second, third and fourth letters in the alphabetical order of the Greek alphabet, respectively.

Nitrile

A Nitrile is any organic compound which has a -C≡N functional group. The -C≡N functional group is called a Nitrile group. In the -CN group, the carbon atom and the nitrogen atom are triple bonded together.

Acetone

Acetone is the organic compound with the formula $OC(CH_3)_2$. This colorless, mobile, flammable liquid is the simplest example of the ketones. Owing to the fact that Acetone is miscible with water, and it serves as an important solvent in its own right, typically the solvent of choice for cleaning purposes in the laboratory.

Halide

A Halide is a binary compound, of which one part is a halogen atom and the other part is an element or radical that is less electronegative than the halogen, to make a fluoride, chloride, bromide, iodide, or astatide compound. Many salts are Halide s. All Group 1 metals form Halide s with the halogens and they are white solids.

Acid anhydride

An Acid anhydride is an organic compound that has two acyl groups bound to the same oxygen atom. Most commonly, the acyl groups are derived from the same carboxylic acid, the formula of the anhydride being $(RC(O))_2O$. Symmetrical Acid anhydride s of this type are named by replacing the word acid in the name of the parent carboxylic acid by the word anhydride. Thus, $(CH_3CO)_2O$ is called acetic anhydride.

Acyl

An Acyl group (IUPAC name: alkanoyl) is a functional group derived by the removal of one or more hydroxyl groups from an oxoacid. In organic chemistry, the Acyl group is usually derived from a carboxylic acid. It therefore has the formula RC (=O)-, where R represents an alkyl group; there is a double bond between the carbon and oxygen atoms, and a single bond between R and the carbon.

Nucleophilic acyl substitution

Nucleophilic acyl substitution describes the substitution reaction involving nucleophiles and acyl compounds. Acyl compounds are carboxylic acid derivatives including esters, amides and acid halides. Nucleophiles include anionic reagents such as alkoxide compounds and enolates or species of high basicity, such as amines .

Hofmann rearrangement

The Hofmann rearrangement is the organic reaction of a primary amide to a primary amine with one fewer carbon atom.

The reaction of bromine with sodium hydroxide forms sodium hypobromite in situ, which transforms the primary amide into an intermediate isocyanate. The intermediate isocyanate is hydrolyzed to a primary amine giving off carbon dioxide.

Esters	Esters are chemical compounds derived formally from an oxoacid (one containing an oxo group, X=O), and a hydroxyl compound such as an alcohol or phenol. Esters consist of an inorganic acid or organic acid in which at least one -OH (hydroxyl) group is replaced by an -O-alkyl (alkoxy) group. They are analogous to salts, using organic alcohols instead of metallic hydroxides.
Transesterification	In organic chemistry, Transesterification is the process of exchanging the organic group R' of an ester with the organic group R' of an alcohol. These reactions are often catalyzed by the addition of an acid or base. Transesterification: alcohol + ester → different alcohol + different ester Acids can catalyse the reaction by donating a proton to the carbonyl group, thus making it more reactive, while bases can catalyse the reaction by removing a proton from the alcohol, thus making it more reactive.
Aspirin	Aspirin is a salicylate drug, often used as an analgesic to relieve minor aches and pains, as an antipyretic to reduce fever, and as an anti-inflammatory medication. Aspirin also has an antiplatelet effect by inhibiting thromboxane prostaglandins, which under normal circumstances bind platelet molecules together to repair damaged blood vessels. This is why Aspirin is used in long-term, low doses to prevent heart attacks, strokes, and blood clot formation in people at high risk for developing blood clots.
Alcohol	In chemistry, an Alcohol is any organic compound in which a hydroxyl group (-OH) is bound to a carbon atom of an alkyl or substituted alkyl group. The general formula for a simple acyclic Alcohol is $C_nH_{2n+1}OH$. In common terms, the word Alcohol refers to ethanol, the type of Alcohol found in Alcohol ic beverages. Ethanol is a colorless, volatile liquid with a mild odor which can be obtained by the fermentation of sugars.
Amine	Amine s are organic compounds and functional groups that contain a basic nitrogen atom with a lone pair. Amine s are derivatives of ammonia, wherein one or more hydrogen atoms have been replaced by a substituent such as an alkyl or aryl group. Important Amine s include amino acids, biogenic Amine s, trimethyl Amine , and aniline; see Category: Amine s for a list of Amine s.
Ketone	In organic chemistry, a Ketone is a type of compound that features a carbonyl group bonded to two other carbon atoms, i.e., $R_3CCO\text{-}CR_3$ where R can be a variety of atoms and groups of atoms. With carbonyl carbon bonded to two carbon atoms, Ketone s are distinct from many other functional groups, such as carboxylic acids, aldehydes, esters, amides, and other oxygen-containing compounds. The double-bond of the carbonyl group distinguishes Ketone s from alcohols and ethers.
Condensation reaction	A Condensation reaction is a chemical reaction in which two molecules or moieties (functional groups) combine to form one single molecule, together with the loss of a small molecule. When this small molecule is water, it is known as a dehydration reaction; other possible small molecules lost are hydrogen chloride, methanol, or acetic acid. The word 'condensation' suggests a process in which something is lost; for reactions a small molecule is lost.
Reagent	A Reagent or reactant is a substance or compound consumed during a chemical reaction. Solvents and catalysts, although they are involved in the reaction, are usually not referred to as reactants.

Although the terms reactant and Reagent are often used interchangeably, a Reagent is more specifically 'a test substance that is added to a system in order to bring about a reaction or to see whether a reaction occurs'.

Substitution reaction

In a Substitution reaction, a functional group in a particular chemical compound is replaced by another group . In organic chemistry, the electrophilic and nucleophilic substitution reactions are of prime importance. Organic substitution reactions are classified in several main organic reaction types depending on whether the reagent that brings about the substitution is considered an electrophile or a nucleophile, whether a reactive intermediate involved in the reaction is a carbocation, a carbanion or a free radical or whether the substrate is aliphatic or aromatic.

Acylation

In chemistry, Acylation is the process of adding an acyl group to a compound. The compound providing the acyl group is called the acylating agent.

Because they form a strong electrophile when treated with some metal catalysts, acyl halides are commonly used as acylating agents.

Cantharidin

Cantharidin, a type of terpenoid, is a poisonous chemical compound secreted by many species of blister beetle, and most notably by the Spanish fly, Lytta vesicatoria. The false blister beetles and cardinal beetles also have Cantharidin. Black Blister Beetle Epicauta pennsylvanica

Cantharidin was first isolated by Pierre Robiquet in 1810.

Coumarin

Coumarin is a chemical compound (benzopyrone); a toxin found in many plants, notably in high concentration in the tonka bean (Dipteryx odorata), vanilla grass (Anthoxanthum odoratum), woodruff (Galium odoratum), mullein (Verbascum spp.), and sweet grass (Hierochloe odorata.) It has a sweet scent, readily recognised as the scent of newly-mown hay, and has been used in perfumes since 1882. It has clinical medical value as the precursor for several anticoagulants, notably warfarin, and is used as a gain medium in some dye lasers.

Polyester

Polyester is a category of polymers which contain the ester functional group in their main chain. Although there are many polyesters, the term 'Polyester' as a specific material most commonly refers to polyethylene terephthalate (PET.) Polyesters include naturally-occurring chemicals, such as in the cutin of plant cuticles, as well as synthetics through step-growth polymerization such as polycarbonate and polybutyrate.

Nylon

Nylon is a generic designation for a family of synthetic polymers known generically as polyamides and first produced on February 28, 1935 by Wallace Carothers at DuPont. Nylon is one of the most commonly used polymers.

Nylon is a thermoplastic silky material, first used commercially in a Nylon-bristled toothbrush (1938), followed more famously by women's stockings ('nylons'; 1940.)

Polyamide

A Polyamide is a polymer containing monomers of amides joined by peptide bonds. They can occur both naturally, examples being proteins, such as wool and silk, and can be made artificially through step-growth polymerization, examples being nylons, aramids, and sodium poly(aspartate.)

The amide link is produced from the condensation reaction of an amino group and a carboxylic acid or acid chloride group.

Thioester	Thioesters are compounds resulting from the bonding of sulfur with an acyl group with the general formula R-S-CO-R'. They are the product of esterification between a carboxylic acid and a thiol (as opposed to an alcohol in regular esters.) In biochemistry a Thioester connects the acetyl groups in acetyl-CoA and malonyl-CoA. Some biochemists believe that the Thioester bond was critical for the origin of life.
Carbamates	Carbamates, or urethanes, are a group of organic compounds sharing a common functional group with the general structure R^1-O-(CO)NR^2-R^3. Carbamates are esters of carbamic acid, NH_2COOH, an unstable compound with $R^1 = R^2 = R^3 = H$. Carbamic acid is an unstable compound because of the lability of the H-groups, leading upon exposure to water to rearrangements that lead to the decompositon to ammonium bicarbonate. Since carbamic acid contains a nitrogen attached to a carbonyl group, it is also an amide.
Carbamic acid	Carbamic acid is a compound that is unstable under normal circumstances. Its importance is due more to its relevance in identifying the names of larger compounds. The radical is called 'carbamoyl'.
Carbonate ester	A Carbonate ester is a functional group in organic chemistry consisting of a carbonyl group flanked by two alkoxy groups. The general structure of these carbonates is $R_1O(C{=}O)OR_2$ and they are related to esters $R_1O(C{=}O)R$ and ethers R_1OR_2 and also to the inorganic carbonates. Carbonate ester s are used as protecting group for diols.
Isocyanate	Isocyanate is the functional group of atoms -N=C=O (1 nitrogen, 1 carbon, 1 oxygen), not to be confused with the cyanate functional group which is arranged as -O-C≡N. Any organic compound which contains an Isocyanate group may also be referred to in brief as an Isocyanate An Isocyanate may have more than one Isocyanate group. An Isocyanate that has two Isocyanate groups is known as a di Isocyanate .
Physostigmine	Physostigmine is a parasympathomimetic, specifically, a reversible cholinesterase inhibitor alkaloid of the Calabar bean. The chemical was synthesized for the first time in 1935 by the chemists Percy Lavon Julian and Josef Pikl. It is available in the U.S. under the trade names Antilirium, Eserine Salicylate, Isopto Eserine, and Eserine Sulfate.
Polycarbonates	Polycarbonates are a particular group of thermoplastic polymers. They are easily worked, moulded, and thermoformed; as such, these plastics are very widely used in the modern chemical industry. Their interesting features (temperature resistance, impact resistance and optical properties) position them between commodity plastics and engineering plastics.
Polyurethane	A Polyurethane, commonly abbreviated PU, is any polymer consisting of a chain of organic units joined by urethane (carbamate) links. Polyurethane polymers are formed through step-growth polymerization by reacting a monomer containing at least two isocyanate functional groups with another monomer containing at least two hydroxyl (alcohol) groups in the presence of a catalyst.

Polyurethane formulations cover an extremely wide range of stiffness, hardness, and densities.

Alpha carbon	The alpha carbon in organic chemistry refers to the first carbon that attaches to a functional group (the carbon is attached at the first position.) By extension, the second carbon is the beta carbon, and so on. This nomenclature can also be applied to the hydrogen atoms attached to the carbons.
Carbon	Carbon is the chemical element with symbol C and atomic number 6. As a member of group 14 on the periodic table, it is nonmetallic and tetravalent--making four electrons available to form covalent chemical bonds. There are three naturally occurring isotopes, with ^{12}C and ^{13}C being stable, while ^{14}C is radioactive, decaying with a half-life of about 5730 years.
Claisen condensation	The Claisen condensation is a carbon-carbon bond forming reaction that occurs between two esters or one ester and another carbonyl compound in the presence of a strong base, resulting in a β-keto ester or a β-diketone. It is named after Rainer Ludwig Claisen, who first published his work on the reaction in 1881 . At least one of the reagents must be enolizable (have an α-proton and be able to undergo deprotonation to form the enolate anion.)
Enol	Enol s are alkenes with a hydroxyl group affixed to one of the carbon atoms composing the double bond. Enol s and carbonyl compounds are in fact isomers; this is called keto Enol tautomerism: The Enol form is shown above. It is usually unstable, does not survive long, and changes into the keto form shown on the right.
Fentanyl	Fentanyl -- brand names include Actiq, Duragesic, Fentora, and Sublimaze -- is a synthetic primary μ-opioid agonist commonly used to treat post-operative and chronic breakthrough pain. It is approximately 100 times more potent than morphine, with 100 micrograms of Fentanyl equivalent to 10 mg of morphine and 75 mg of pethidine in analgesic activity. It has an LD_{50} of 3.1 milligrams per kilogram in rats, 0.03 milligrams per kilogram in monkeys, and an undetermined milligrams per kilogram in humans.
Keto-enol tautomerism	In organic chemistry, Keto-enol tautomerism refers to a chemical equilibrium between a keto form (a ketone or an aldehyde) and an enol. The enol and keto forms are said to be tautomers of each other. The interconversion of the two forms involves the movement of a proton and the shifting of bonding electrons; hence, the isomerism qualifies as tautomerism.
Ketone	In organic chemistry, a Ketone is a type of compound that features a carbonyl group bonded to two other carbon atoms, i.e., $R_3CCO\text{-}CR_3$ where R can be a variety of atoms and groups of atoms. With carbonyl carbon bonded to two carbon atoms, Ketone s are distinct from many other functional groups, such as carboxylic acids, aldehydes, esters, amides, and other oxygen-containing compounds. The double-bond of the carbonyl group distinguishes Ketone s from alcohols and ethers.
Hydrogen	Hydrogen is the chemical element with atomic number 1. It is represented by the symbol H. At standard temperature and pressure, Hydrogen is a colorless, odorless, nonmetallic, tasteless, highly flammable diatomic gas with the molecular formula H_2. With an atomic weight of 1.007 94 u, Hydrogen is the lightest element.
Diisopropylamine	Diisopropylamine is a secondary amine with the chemical formula $(CH_3)_2HC\text{-}NH\text{-}CH(CH_3)_2$. It is best known as its lithium salt, lithium diisopropylamide, known as 'LDA'. LDA is a strong, non-nucleophilic base.

Lithium diisopropylamide

Lithium diisopropylamide is the chemical compound with the formula $[(CH_3)_2CH]_2NLi$. Generally abbreviated Lithium diisopropylamide A, it is a strong base used in organic chemistry for the deprotonation of weakly acidic compounds. The reagent has been widely accepted because it is soluble in non-polar organic solvents and it is non-pyrophoric.

Halogenation

Halogenation is a chemical reaction that incorporates a halogen atom into a molecule. More specific descriptions exist that specify the type of halogen: fluorination, chlorination, bromination, and iodination.

In a Markovnikov addition reaction, a halogen like bromine is reacted with an alkene which causes the π-bond to break forming an haloalkane.

Alkanes

Alkanes are chemical compounds that consist only of the elements carbon and hydrogen, wherein these atoms are linked together exclusively by single bonds without any cyclic structure Alkanes belong to a homologous series of organic compounds in which the members differ by a constant relative atomic mass of 14.

Each carbon atom must have 4 bonds, and each hydrogen atom must be joined to a carbon atom

Haloform reaction

The Haloform reaction is a chemical reaction where a haloform (CHX_3, where X is a halogen) is produced by the exhaustive halogenation of a methyl ketone (a molecule containing the $R\text{-}CO\text{-}CH_3$ group) in the presence of a base. R may be H, alkyl or aryl. The reaction can be used to produce $CHCl_3$, $CHBr_3$ or CHI_3.

Iodoform

Iodoform is the organoiodine compound with the formula CHI_3. A pale yellow, crystalline, volatile substance, it has a penetrating odor (in older chemistry texts, the smell is sometimes referred to as the smell of hospitals) and, analogous to chloroform, sweetish taste. It is occasionally used as a disinfectant.

Acetone

Acetone is the organic compound with the formula $OC(CH_3)_2$. This colorless, mobile, flammable liquid is the simplest example of the ketones. Owing to the fact that Acetone is miscible with water, and it serves as an important solvent in its own right, typically the solvent of choice for cleaning purposes in the laboratory.

Acetophenone

Acetophenone is the organic compound with the formula $C_6H_5C(O)CH_3$. It is the simplest aromatic ketone. This colourless, viscous liquid is a precursor to useful resins and fragrances.

Carboxylic acids

Carboxylic acids are organic acids characterized by the presence of a carboxyl group, which has the formula -C(=O)OH, usually written -COOH or $-CO_2H$. Carboxylic acids are Brønsted-Lowry acids -- they are proton donors. Salts and anions of Carboxylic acids are called carboxylates.

The simplest series of Carboxylic acids are the alkanoic acids, R-COOH, where R is a hydrogen or an alkyl group.

Alkynes

Alkynes are hydrocarbons that have a triple bond between two carbon atoms, with the formula C_nH_{2n-2}. Alkynes are traditionally known as acetylenes, although the name acetylene also refers specifically to C_2H_2, known formally (but rarely) as ethyne using IUPAC nomenclature. Like other hydrocarbons, Alkynes are generally hydrophobic but tend to be more reactive.

Alkylation	Alkylation is the transfer of an alkyl group from one molecule to another. The alkyl group may be transferred as an alkyl carbocation, a free radical, a carbanion or a carbene (or their equivalents) . Alkylating agents are widely used in chemistry because the alkyl group is probably the most common group encountered in organic molecules.
Enamine	An Enamine is an unsaturated compound derived by the reaction of an aldehyde or ketone with a secondary amine followed by loss of H_2O. The word 'Enamine' is derived from the prefix en-, used as the suffix of alkene, and the root amine. Compare with enol, which is a molecule containing both alkene (en-) and alcohol (-ol.) If one of the nitrogen substituents is H, it is the tautomeric form of an imine.
Acylation	In chemistry, Acylation is the process of adding an acyl group to a compound. The compound providing the acyl group is called the acylating agent. Because they form a strong electrophile when treated with some metal catalysts, acyl halides are commonly used as acylating agents.
Aldol	An Aldol or Aldol adduct is a beta-hydroxy ketone or aldehyde, and is the product of Aldol addition (as opposed to Aldol condensation, which produces an α,β-unsaturated carbonyl moiety.) Typically, 'Aldol' refers to 3-hydroxybutanal.
Aldol condensation	An Aldol condensation is an organic reaction in which an enolate ion reacts with a carbonyl compound to form a β-hydroxyaldehyde or β-hydroxyketone, followed by dehydration to give a conjugated enone. Aldol condensation s are important in organic synthesis, providing a good way to form carbon-carbon bonds. The Robinson annulation reaction sequence features an Aldol condensation the Wieland-Miescher ketone product is an important starting material for many organic syntheses.
Acetaldehyde	Acetaldehyde is an organic chemical compound with the formula CH_3CHO or MeCHO. It is a flammable liquid. Acetaldehyde occurs naturally in ripe fruit, coffee, and bread, and is produced by plants as part of their normal metabolism. It is popularly known as a chemical that causes hangovers.
Diacetone alcohol	Diacetone alcohol is a chemical compound with the formula $CH_3C(O)CH_2C(OH)(CH_3)_2$. This liquid is common synthetic intermediate used for the preparation of other compounds. It is a natural ingredient of Sleepy grass (Achnatherum robustum.)
2,3-Dimethylbutane	2,3-Dimethylbutane is an isomer of hexane. It has the chemical formula$_2CHCH_2$ It is a colourless liquid which boils at 57.9°C. .
Diisopropyl ether	Diisopropyl ether is secondary ether that is used as a solvent. It is a colorless liquid that is slightly soluble in water, but miscible with most organic solvents. It is also used as an oxygenate gasoline additive.

Esters	Esters are chemical compounds derived formally from an oxoacid (one containing an oxo group, X=O), and a hydroxyl compound such as an alcohol or phenol. Esters consist of an inorganic acid or organic acid in which at least one -OH (hydroxyl) group is replaced by an -O-alkyl (alkoxy) group. They are analogous to salts, using organic alcohols instead of metallic hydroxides.
Ethyl acetate	Ethyl acetate is the organic compound with the formula $CH_3COOCH_2CH_3$. This colorless liquid has a characteristic sweet smell (similar to pear drops) like certain glues or nail polish removers, in which it is used. Ethyl acetate is the ester of ethanol and acetic acid; it is manufactured on a large scale for use as a solvent.
Petroleum	Petroleum or crude oil is a naturally occurring, flammable liquid found in rock formations in the Earth consisting of a complex mixture of hydrocarbons of various molecular weights, plus other organic compounds. The term 'Petroleum' was first used in the treatise De Natura Fossilium, published in 1546 by the German mineralogist Georg Bauer, also known as Georgius Agricola. The proportion of hydrocarbons in the mixture is highly variable and ranges from as much as 97% by weight in the lighter oils to as little as 50% in the heavier oils and bitumens.
Fatty acid	In chemistry, especially biochemistry, a Fatty acid is a carboxylic acid often with a long unbranched aliphatic tail (chain), which is either saturated or unsaturated. Carboxylic acids as short as butyric acid (4 carbon atoms) are considered to be Fatty acid s, whereas Fatty acid s derived from natural fats and oils may be assumed to have at least eight carbon atoms, caprylic acid (octanoic acid), for example. The most abundant natural Fatty acid s have an even number of carbon atoms because their biosynthesis involves acetyl-CoA, a coenzyme carrying a two-carbon-atom group
Fatty acids	Fatty acids are an important source of energy for many organisms. Excess glucose can be stored efficiently as fat. Triglycerides yield more than twice as much energy for the same mass as do carbohydrates or proteins.
Methyl formate	Methyl formate is the methyl ester of formic acid. It is a clear liquid with an ethereal odor, high vapor pressure and low surface tension. In the laboratory, Methyl formate can be produced by the condensation reaction of methanol and formic acid, as follows: $HCOOH + CH_3OH \rightarrow HCOOCH_3 + H_2O$ Industrial Methyl formate, however, is usually produced by the combination of methanol and carbon monoxide in the presence of a strong base, such as sodium methoxide: $CH_3OH + CO \rightarrow HCOOCH_3$ This process, practised commercially by BASF amongst other companies gives 96 % selectivity towards Methyl formate, although can suffer from catalyst sensitivity to water which can commonly be present in the carbon monoxide feedstock, commonly derived from synthesis gas.

Methylene	Methylene is a chemical species in which a carbon atom is bonded to two hydrogen atoms. Three different possibilities present themselves: • the $-CH_2-$ group, e.g. dichloromethane (also known as Methylene chloride) • the $=CH_2$ group, e.g. methylenecyclopropene, • and the $:CH_2$ molecule, a carbene known as Methylene. A carbene is a highly reactive organic molecule having a divalent carbon atom with six valence electrons. Methylene groups in a chain or ring contribute to its size and lipophilicity.
Isomer	In chemistry, Isomer s are compounds with the same molecular formula but different structural formula. Isomer s do not necessarily share similar properties unless they also have the same functional groups. This should not be confused with a nuclear Isomer which involves a nucleus at different states of excitement.
Acetic acid	Acetic acid, CH_3COOH, also known as ethanoic acid, is an organic acid which gives vinegar its sour taste and pungent smell. Pure, water-free Acetic acid is a colourless liquid that absorbs water from the environment (hygroscopy), and freezes at 16.7 °C (62 °F) to a colourless crystalline solid. It is a weak acid, in that it is only partially dissociated acid in aqueous solution.
Diethyl malonate	Diethyl malonate is the diethyl ester of malonic acid. It occurs naturally in grapes and strawberries as a colourless liquid with an apple-like odour, and is used in perfumes. It is also used to synthesize other compounds such as barbiturates, artificial flavourings, vitamin B1, and vitamin B6.
Malonic ester synthesis	The Malonic ester synthesis is a chemical reaction where diethyl malonate or another ester of malonic acid is alkylated at the carbon alpha (directly adjacent) to both carbonyl groups, and then converted to a substituted acetic acid. A strong base is required to deprotonate the center proton. The protons alpha to carbonyl groups are easily deprotonated.
Biosynthesis	Biosynthesis is a phenomenon wherein chemical compounds are produced from simpler reagents. Biosynthesis, unlike chemosynthesis, takes place within living organisms and is generally catalyzed by enzymes. The process is a vital part of metabolism.
Acetoacetic ester synthesis	Acetoacetic ester synthesis is a chemical reaction where ethyl acetoacetate is alkylated at the α-carbon to both carbonyl groups and then converted into a ketone, or more specifically an α-substituted acetone. This is very similar to malonic ester synthesis. A strong base deprotonates the dicarbonyl α-carbon.

Ethyl acetoacetate	The organic compound Ethyl acetoacetate is the ethyl ester of acetoacetic acid. It is a solvent, and is used as a chemical intermediate in the production of a wide variety of compounds, such as amino acids, analgesics, antibiotics, antimalarial agents, antipyrine and aminopyrine, and vitamin B1; and also in the manufacture of dyes, inks, lacquers, perfumes, plastics, and yellow paint pigments. Alone, it is used as a flavoring for food.
Electrophilic addition	In organic chemistry, an Electrophilic addition reaction is an addition reaction where, in a chemical compound, a pi bond is removed by the creation of two new covalent bonds. The substrate of an Electrophilic addition reaction must have a double bond or triple bond . The driving force for this reaction is the formation of an electrophile X^+ that forms a covalent bond with an electron-rich unsaturated C=C bond.
Michael reaction	The Michael reaction or Michael addition is the nucleophilic addition of a carbanion to an alpha, beta unsaturated carbonyl compound. It belongs to the larger class of conjugate additions. This is one of the most useful methods for the mild formation of C-C bonds.
Robinson annulation	The Robinson annulation is an organic reaction used to create a six-member ring α,β-unsaturated cyclic ketone, using a ketone (or aldehyde) and methyl vinyl ketone. It is named after Sir Robert Robinson, the British chemist who discovered it while he was at the University of Oxford. In addition to methyl vinyl ketone, 1-chloro-3-butanone and isoxazoles will give the same product.

Carbohydrates

Carbohydrates or saccharides are the most abundant of the four major classes of biomolecules. They fill numerous roles in living things, such as the storage and transport of energy (eg: starch, glycogen) and structural components (eg: cellulose in plants and chitin.) Additionally, Carbohydrates and their derivatives play major roles in the working process of the immune system, fertilization, pathogenesis, blood clotting, and development.

Cellulose

Cellulose is an organic compound with the formula $(C_6H_{10}O_5)$Template:Chem/dispAAA, a polysaccharide consisting of a linear chain of several hundred to over ten thousand β(1→4) linked D-glucose units.

Cellulose is the structural component of the primary cell wall of green plants, many forms of algae and the oomycetes. Some species of bacteria secrete it to form biofilms.

Cellulose acetate

Cellulose acetate, first prepared in 1865, is the acetate ester of cellulose. Cellulose acetate is used as a film base in photography, and as a component in some adhesives; it is also used as a synthetic fiber.

Acetate and triacetate are mistakenly referred to as the same fiber; although they are similar, their chemical compounds differ.

Glycogen

Glycogen is the molecule which functions as the secondary short term energy storage in animal cells. It is made primarily by the liver and the muscles, but can also be made by Glycogen esis within the brain and stomach. Glycogen is the analogue of starch, a less branched glucose polymer in plants, and is commonly referred to as animal starch, having a similar structure to amylopectin.

Rayon

Rayon is a manufactured regenerated cellulose fiber. Because it is produced from naturally occurring polymers, it is neither a truly synthetic fiber nor a natural fiber, it is a semi-synthetic fiber. Rayon is known by the names viscose Rayon and art silk in the textile industry.

Starch

Starch or amylum is a polysaccharide carbohydrate consisting of a large number of glucose units joined together by glycosidic bonds. Starch is produced by all green plants as an energy store and is a major food source for humans.

Pure Starch is a white, tasteless and odorless powder that is insoluble in cold water or alcohol.

Aldose

An Aldose is a monosaccharide (a simple sugar) containing one aldehyde group per molecule and having a chemical formula of the form $C_n(H_2O)_n$.

With only two carbon atoms, glycolaldehyde is the simplest of all Aldose s.

Aldose s isomerize to ketoses in the Lobry-de Bruyn-van Ekenstein transformation.

Biopolymer

Biopolymer s are a class of polymers produced by living organisms. Cellulose and starch, proteins and peptides, and DNA and RNA are all examples of Biopolymer s, in which the monomeric units, respectively, are sugars, amino acids, and nucleotides.

Cellulose is the most common Biopolymer and the common organic compound on Earth, and about 33 percent of all plant matter is cellulose (the cellulose content of cotton is 90 percent and that of wood is 50 percent)

	A major but defining difference between polymers and Biopolymer s can be found in their structures.
Dipropylamine	Dipropylamine is a flammable, highly toxic, corrosive amine. It occurs naturally in tobacco leaves and artificially in industrial wastes. Exposure can cause excitement followed by depression, internal bleeding, dystrophy, and severe irritation.
Disaccharide	A Disaccharide is the carbohydrate formed when two monosaccharides undergo a condensation reaction which involves the elimination of a small molecule, such as water, from the functional groups only. Like monosaccharides, Disaccharide s also dissolve in water, taste sweet and are called sugars. Disaccharide is one of the four chemical groupings of carbohydrates (monosaccharide, Disaccharide oligosaccharide, and polysaccharide.)
Fischer projection	The Fischer projection, devised by Hermann Emil Fischer in 1891, is a two-dimensional representation of a three-dimensional organic molecule by projection. They are used by chemists, particularly in organic chemistry and biochemistry. All bonds are depicted as horizontal or vertical lines.
Fructose	Fructose is a simple reducing sugar found in many foods and is one of the three important dietary monosaccharides along with glucose and galactose. Honey, tree fruits, berries, melons, and some root vegetables, such as beets, sweet potatoes, parsnips, and onions, contain Fructose, usually in combination with glucose in the form of sucrose. Fructose is also derived from the digestion of granulated table sugar (sucrose), a disaccharide consisting of glucose and Fructose.
Glucose	Glucose, a monosaccharide also known as grape sugar, blood sugar is a very important carbohydrate in biology. The living cell uses it as a source of energy and metabolic intermediate. Glucose is one of the main products of photosynthesis and starts cellular respiration in both prokaryotes and eukaryotes
Ketose	A Ketose is a sugar containing one ketone group per molecule. With 3 carbon atoms, dihydroxyacetone is the simplest of all Ketose s and is the only one having no optical activity. Ketose s can isomerize into an aldose when the carbonyl group is located at the end of the molecule.
Monosaccharides	Monosaccharides are the most basic unit of carbohydrates. They are the simplest form of sugar and are usually colorless, water-soluble, crystalline solids. Some Monosaccharides have a sweet taste.
Sucrose	Sucrose is a disaccharide of glucose and fructose with an α 1,2 glycosidic linkage. The molecular formula of Sucrose is $C_{12}H_{22}O_{11}$. Its systematic name is β-D-fructofuranosyl--α-D-glucopyranoside
Sugar	Sugar is a class of edible crystalline substances, mainly sucrose, lactose, and fructose. Human taste buds interpret its flavor as sweet. Sugar as a basic food carbohydrate primarily comes from Sugar cane and from Sugar beet, but also appears in fruit, honey, sorghum, Sugar maple (in maple syrup), and in many other sources.
Aldohexose	An Aldohexose is a hexose with an aldehyde group on one end.

The Aldohexose s have four chiral centres for a total of 16 possible Aldohexose stereoisomers (2^4.) Of these, only three commonly occur in nature: D-glucose, D-galactose, and D-mannose.

Ketohexose

A Ketohexose is a ketone-containing hexose (a six-carbon monosaccharide.) Ketohexose s have three chiral centers, so 8 ($2^3 = 8$) different stereoisomers are possible.

The 4 D- Ketohexose s are:

CH_2OH
CH_2OH
CH_2OH
CH_2OH
| | | |
C=O
C=O
C=O
C=O | |
| | HC-
OH HO-
CH HC-
OH HO-
CH | | |
| HC-OH
HC-OH
HO-CH
HO-CH |
| | | HC-
OH HC-
OH HC-
OH HC-
OH | | |
|
CH_2OH
CH_2OH
CH_2OH
CH_2OH
D-
Psicose
D-
Fructose
D-
Sorbose
D-
Tagatose

Multiple

Major families of biochemicalsSaccharides/Carbohydrates/Glycosides Â· Amino acids/Peptides/Proteins/Glycoproteins Â· Lipids/Terpenes/Steroids/Carotenoids Â· Alkaloids/Nucleobases/Nucleic acids Â· Cofactors/Flavonoids/Polyketides/Tetrapyrroles

Asymmetric carbon	The term Asymmetric carbon is an expression used, mainly in older literature, for what is now more commonly called a stereogenic carbon. This expression refers to a carbon atom that is attached to four different atoms or four different groups of atoms. Knowing the number of asymmetric (stereogenic) carbon atoms, one can calculate the maximum possible number of enantiomers for any given molecule as follows:

If n is the number of stereogenic carbon atoms then the maximum number of isomers = 2^n

As an example, the molecule shown here has a stereogenic carbon represented by the C in the center of the structure: .

Configuration

The configuration of a molecule is the permanent geometry that results from the spatial arrangement of its bonds. The ability of the same set of atoms to form two or more molecules with different configurations is stereoisomerism. configuration is distinct from chemical conformation, a shape attainable by bond rotations.

Diastereomers

Diastereomers are stereoisomers that are not enantiomers (non-superimposable mirror images of each other.) Diastereomers can have different physical properties and different reactivity. In another definition Diastereomers are pairs of isomers that have opposite configurations at one or more of the chiral centers but are not mirror images of each other .

Epimers

In chemistry, Epimers are diastereomers that differ in configuration of only one stereogenic center. Diastereomers are a class of stereoisomers that are non-superposable, non-mirror images of one another, unlike enantiomers which are non-superposable mirror images of one another.

For example, the sugars α-glucose and β-glucose are Epimers.

Hemiacetals

Hemiacetals and hemiketals are compounds of the general formula $R_1R'_1C(OH)OR_2$, where R_2 is not hydrogen. These types of compounds are formed from carbonyl compounds and alcohols, i.e. Hemiacetals from aldehydes, hemiketals from ketones. Thus, for Hemiacetals, either R_1 or R'_1 must be a hydrogen, while for hemiketals, neither will be.

Haworth projection

A Haworth projection is a common way of representing the cyclic structure of monosaccharides with a simple three-dimensional perspective.

The Haworth projection was named after the English chemist Sir Walter N. Haworth.

A Haworth projection has the following characteristics:

- Carbon is the implicit type of atom. In the example on the right, the atoms numbered from 1 to 6 are all carbon atoms. Carbon 1 is known as the Anomeric Carbon.
- Hydrogen atoms on carbon are implicit. In the example, atoms 1 to 6 have extra hydrogen atoms not depicted.
- A thicker line indicates atoms that are closer to the observer. In the example on the right, atoms 2 and 3 are the closest to the observer, atoms 1 and 4 are further from the observer and finally the remaining atoms (5, etc.) are the furthest.

.

Pyranose

Pyranose is a collective term for carbohydrates which have a chemical structure that includes a six-membered ring consisting of five carbon atoms and one oxygen atom. The name derives from its similarity to the oxygen heterocycle pyran.

The Pyranose ring is formed by the reaction of the hydroxyl group on carbon 5 (C-5) of a sugar with the aldehyde at carbon 1.

Carbon

Carbon is the chemical element with symbol C and atomic number 6. As a member of group 14 on the periodic table, it is nonmetallic and tetravalent--making four electrons available to form covalent chemical bonds. There are three naturally occurring isotopes, with ^{12}C and ^{13}C being stable, while ^{14}C is radioactive, decaying with a half-life of about 5730 years.

Epoxide

An Epoxide is a cyclic ether with three ring atoms. This ring approximately defines an equilateral triangle, which makes it highly strained. The strained ring makes Epoxide s more reactive than other ethers.

Furan

Furan is a heterocyclic organic compound. It is typically derived by the thermal decomposition of pentose-containing materials, cellulosic solids especially pine-wood. Furan is a colorless, flammable, highly volatile liquid with a boiling point close to room temperature.

Tetrahydropyran

Tetrahydropyran is the organic compound consisting of a saturated six-membered ring containing five carbon atoms and one oxygen atom.

The Tetrahydropyran ring system is the core of pyranose sugars.

One classic procedure for the organic synthesis of Tetrahydropyran is by hydrogenation with Raney nickel of dihydropyran.

Phenol

In organic chemistry, phenols, sometimes called phenolics, are a class of chemical compounds consisting of a hydroxyl group (-OH) bonded directly to an aromatic hydrocarbon group. The simplest of the class is Phenol Phenol - the simplest of the phenols.

Although similar to alcohols, phenols have unique properties and are not classified as alcohols (since the hydroxyl group is not bonded to a saturated carbon atom.)

Pyran

In chemistry, a Pyran is a six membered heterocyclic ring consisting of five carbon atoms and one oxygen atom and containing two double bonds. The molecular formula is C_5H_6O. There are two isomers of Pyran that differ by the location of the double bonds. In 2H-Pyran, the saturated carbon is at position 2, whereas in 4H-Pyran, the saturated carbon is at position 4.

Anomer

In carbohydrate chemistry, an Anomer is a special type of epimer. It is a stereoisomer (diastereomer, more exactly) of a cyclic saccharide that differs only in its configuration at the hemiacetal (or hemiketal) carbon, also called the Anomer ic carbon.

Anomer s are identified as 'α' or 'β' based on the relation between the stereochemistry of the Anomer ic carbon and the furthest chiral centre in the ring.

Mutarotation	Mutarotation is the term given to the change in the specific rotation of a cyclic monosaccharide as it reaches an equilibrium between its α and β anomeric forms. It was discovered by Dubrunfaut in 1846, when he noticed that the specific rotation of aqueous sugar solution changes with time. The optical rotation of the solution depends on the optical rotation of each anomer and their ratio in the solution.
Sorbitol	Sorbitol is a sugar alcohol that the body metabolises slowly. It is obtained by reduction of glucose changing the aldehyde group to an additional hydroxyl group. Sorbitol is a sugar substitute often used in diet foods and sugar-free chewing gum, mints and cough syrups.
Sugar alcohol	A Sugar alcohol is a hydrogenated form of carbohydrate, whose carbonyl group (aldehyde or ketone, reducing sugar) has been reduced to a primary or secondary hydroxyl group (hence the alcohol.) Sugar alcohols have the general formula H $(HCHO)_{n+1}H$, whereas sugars have $H(HCHO)_nHCO$. In commercial foodstuffs sugar alcohols are commonly used in place of table sugar (sucrose), often in combination with high intensity artificial sweeteners to counter the low sweetness. Sugar alcohols do not contribute to tooth decay.
Aldonic acid	An Aldonic acid is any of a family of sugar acids obtained by oxidation of the aldehyde functional group of an aldoses to form a carboxylic acid functional group. Thus, their general chemical formula is $HOOC\text{-}(CHOH)_n\text{-}CH_2OH$. Oxidation of the terminal hydroxyl group instead of the terminal aldehyde yields a uronic acid, while oxidation of both terminal ends yields an aldaric acid. Aldonic acid s are typically prepared by oxidation of the sugar with bromine.
Galactitol	Dulcitol (or Galactitol) is a sugar alcohol, the reduction product of galactose. In people with galactokinase deficiency, a form of galactosemia, excess dulcitol formation in the lens of the eye leads to cataracts. Galactitol is produced from galactose in a reaction catalyzed by aldose reductase.
Mannitol	Mannitol is an organic compound with the formula ($C_6H_8(OH)_6$.) This polyol is used as an osmotic diuretic agent and a weak renal vasodilator. It was originally isolated from the secretions of the flowering ash, called manna after their resemblance to the Biblical food, and is also be referred to as mannite and manna sugar.
Aldaric acid	Aldaric acid is a group of sugar acids characterised by the formula HOOC-(CHOH)n-COOH. Aldaric acid s are usually synthesized by the oxidation of aldoses with nitric acid. In this reaction it is the open-chain (polyhydroxyaldehyde) form of the sugar that reacts. An Aldaric acid is an aldose in which both the hydroxyl function of the terminal carbon and the aldehyde function of the first carbon have been fully oxidized to carboxylic acid functions.
Glucoside	A Glucoside is a glycoside that is derived from glucose. Glucoside s are common in plants, but rare in animals. Glucose is produced when a Glucoside is hydrolysed by purely chemical means, or decomposed by fermentation or enzymes.

Ether	Ether is a class of organic compounds which contain an Ether group -- an oxygen atom connected to two (substituted) alkyl or aryl groups -- of general formula R-O-R'. A typical example is the solvent and anesthetic diethyl Ether, commonly referred to simply as 'Ether' (ethoxyethane, CH_3-CH_2-O-CH_2-CH_3.) Ether molecules cannot form hydrogen bonds amongst each other, resulting in a relatively low boiling point compared to that of the analogous alcohols.
Esters	Esters are chemical compounds derived formally from an oxoacid (one containing an oxo group, X=O), and a hydroxyl compound such as an alcohol or phenol. Esters consist of an inorganic acid or organic acid in which at least one -OH (hydroxyl) group is replaced by an -O-alkyl (alkoxy) group. They are analogous to salts, using organic alcohols instead of metallic hydroxides.
Claisen condensation	The Claisen condensation is a carbon-carbon bond forming reaction that occurs between two esters or one ester and another carbonyl compound in the presence of a strong base, resulting in a β-keto ester or a β-diketone. It is named after Rainer Ludwig Claisen, who first published his work on the reaction in 1881 . At least one of the reagents must be enolizable (have an α-proton and be able to undergo deprotonation to form the enolate anion.)
Glucuronic acid	Glucuronic acid is a carboxylic acid. Its structure is similar to that of glucose. However, Glucuronic acid's sixth carbon is oxidized to a carboxylic acid.
Osazones	Osazones are a class of carbohydrate derivatives found in organic chemistry formed when sugars are reacted with phenylhydrazine. The famous German chemist Emil Fischer developed and used the reaction to identify sugars whose stereochemistry differed by only one chiral carbon. The reaction involves formation of a pair of phenylhydrazone functionalities, concomitant with the oxidation of the hydroxymethylene group adjacent to the formyl center.
Phenylhydrazine	Phenylhydrazine is the chemical compound with the formula $C_6H_5NHNH_2$. Organic chemists abbreviate the compound $PhNHNH_2$. Phenylhydrazine forms monoclinic prisms that melt to an oil around room temperature which may turn yellow to dark red upon exposure to air.
Cyanohydrin	A Cyanohydrin is a functional group found in organic compounds. Cyanohydrin s have the formula $R_2C(OH)CN$, where R is H, alkyl, or aryl. Cyanohydrin s are industrially important precursors to carboxylic acids and some amino acids.
Kiliani-Fischer synthesis	The Kiliani-Fischer synthesis is a method for synthesizing monosaccharides. It proceeds via synthesis and hydrolysis of a cyanohydrin, thus elongating the carbon chain of an aldose by one carbon atom while preserving stereochemistry on all the old chiral carbons. The new chiral carbon is produced with both stereochemistries, so the product of a Kiliani-Fischer synthesis is a mixture of two diastereomeric sugars, called epimers.
Maltose	Maltose is a disaccharide formed from two units of glucose joined with an α(1→4) linkage. It is the second member of an important biochemical series of glucose chains. The addition of another glucose unit yields maltotriose; further additions will produce dextrins (also called maltodextrins) and eventually starch (glucose polymer.)

Cellobiose

Cellobiose is a disaccharide with the formula $[HOCH_2CHO(CHOH)_3]_2O$. The molecule is derived from the condensation of two glucose molecules linked in a β(1→4) bond. It can be hydrolyzed by bacteria or cationic ion exchange resins to give glucose. Cellobiose has eight free alcohol (COH) groups and three ether linkages, which give rise to strong -inter- and intra-molecular hydrogen bonds

It can be obtained by enzymatic or acidic hydrolysis of cellulose and cellulose rich materials such as cotton, jute, or paper.

Lactose

Lactose is a sugar that is found most notably in milk. Lactose makes up around 2-8% of milk (by weight.) It is extracted from sweet or sour whey.

Gentiobiose

Gentiobiose is a disaccharide composed of two units of D-glucose joined with a β(1->6) linkage. It is a white crystalline solid that is soluble in water or hot methanol. Gentiobiose is incorporated into the chemical structure of crocin, the chemical compound that gives saffron its color.

Cellophane

Cellophane is a thin, transparent sheet made of regenerated cellulose. Its low permeability to air, oils and greases, and bacteria makes it useful for food packaging. Cellophane is in many countries a registered trade mark of Innovia Films Ltd, Cumbria, UK. Cellulose is treated with alkali and carbon disulfide to yield viscose.

Cellulose fibers from wood, cotton, hemp, or other sources are dissolved in alkali and carbon disulfide to make a solution called viscose, which is then extruded through a slit into a bath of dilute sulfuric acid and sodium sulfate to reconvert the viscose into cellulose.

Polypropylene

Polypropylene or polypropene (PP) is a thermoplastic polymer, made by the chemical industry and used in a wide variety of applications, including packaging, textiles (e.g. ropes, thermal underwear and carpets), stationery, plastic parts and reusable containers of various types, laboratory equipment, loudspeakers, automotive components, and polymer banknotes. An addition polymer made from the monomer propylene, it is rugged and unusually resistant to many chemical solvents, bases and acids.

In 2007, the global market for Polypropylene had a volume of 45.1 million tons which led to a turnover of about 65 billion US $ (47,4 billion â,¬.)

Polysaccharides

Polysaccharides are polymeric carbohydrate structures, formed of repeating units (either mono- or di-saccharides) joined together by glycosidic bonds. These structures are often linear, but may contain various degrees of branching. Polysaccharides are often quite heterogeneous, containing slight modifications of the repeating unit.

Viscose

Viscose is a viscous organic liquid used to make rayon and cellophane. Viscose is becoming synonymous with rayon, a soft material commonly used in shirt, coats, jackets, and other outer wear.

Cellulose from wood or cotton fibres is treated with sodium hydroxide, then mixed with carbon disulfide to form cellulose xanthate, which is dissolved in more sodium hydroxide.

Amylose

Amylose is a linear polymer of glucose linked mainly by α(1→4) bonds. It can be made of several thousand glucose units. It is one of the two components of starch, the other being amylopectin.

Amyl

The word or root Amyl has two meanings, in organic chemistry and biochemistry.

In biochemistry, Amyl means 'pertaining to starch', ' Amyl um' being another word for starch. Many terms for moderately complex biological chemicals related to starch contain Amyl , for example:

- Amyl ase
- Amyl opectin
- Amyl ose
- Amyl oplast
- Amyl oid (based on an early mistaken opinion that these structures contained starch)

Note that in this usage, it is a part of the word, and becomes ' Amyl o' when preceding a consonant.

In organic chemistry, Amyl is the old trivial name for the alkyl functional group and radical called pentyl under the IUPAC nomenclature: that is, $-C_5H_{11}$.

Amino sugar

In chemistry, an Amino sugar contains an amine group in place of a hydroxyl group. Derivatives of amine containing sugars, such as N-acetylglucosamine and sialic acid, while not formally containing an amine, are also considered Amino sugar s.

Aminoglycosides are a class of antimicrobial compounds that inhibit bacterial protein synthesis.

Chitin

$Chitin_n$ is a long-chain polymer of a N-acetylglucosamine, a derivative of glucose, and is found in many places throughout the natural world. It is the main component of the cell walls of fungi, the exoskeletons of arthropods, such as crustaceans and insects, including ants, beetles and butterflies, the radula of mollusks and the beaks of cephalopods, including squid and octopuses. Chitin has also proven useful for several medical and industrial purposes.

N-acetylglucosamine

N-Acetylglucosamine is a monosaccharide derivative of glucose. Chemically it is an amide between glucosamine and acetic acid. It has a molecular formula of $C_8H_{15}NO_6$, a molar mass of 221.21 g/mol, and it is significant in several biological systems.

Nucleic acid

A Nucleic acid is a macromolecule composed of chains of monomeric nucleotides. In biochemistry these molecules carry genetic information or form structures within cells. The most common nucleic acids are deoxyribonucleic acid (DNucleic acid) and ribonucleic acid (RNucleic acid.)

Nucleotide

Nucleotides are molecules that, when joined together, make up the structural units of RNA and DNA. Additionally, nucleotides play central roles in metabolism. In that capacity, they serve as sources of chemical energy (adenosine triphosphate and guanosine triphosphate), participate in cellular signaling (cyclic guanosine monophosphate and cyclic adenosine monophosphate), and are incorporated into important cofactors of enzymatic reactions (coenzyme A, flavin adenine dinucleotide, flavin mononucleotide, and nicotinamide adenine dinucleotide phosphate.) Figure 1: Structural elements of the most common nucleotides Figure 2: Ribose structure indicating numbering of carbon atoms

A Nucleotide is composed of a nucleobase (nitrogenous base) and a five-carbon sugar (either ribose or 2'-deoxyribose), and one to three phosphate groups.

Uric acid

Uric acid is an organic compound of carbon, nitrogen, oxygen and hydrogen with the formula $C_5H_4N_4O_3$.

Uric acid is produced by xanthine oxidase from xanthine and hypoxanthine, which in turn are produced from purine. Uric acid is more toxic to tissues than either xanthine or hypoxanthine.

Deoxyribose

Deoxyribose, also known as D-Deoxyribose and 2-Deoxyribose, is an aldopentose -- a monosaccharide containing five carbon atoms, and including an aldehyde functional group in its linear structure. It is a deoxy sugar derived from the pentose sugar ribose by the replacement of the hydroxyl group at the 2 position with hydrogen, leading to the net loss of an oxygen atom. Replacement of the hydroxyl group also shifts the conformation of the ring from C3'-endo to C2'-endo.

Primary structure

In biochemistry, the Primary structure of a biological molecule is the exact specification of its atomic composition and the chemical bonds connecting those atoms (including stereochemistry.) For a typical unbranched, un-crosslinked biopolymer (such as a molecule of DNA, RNA or typical intracellular protein), the Primary structure is equivalent to specifying the sequence of its monomeric subunits, e.g., the nucleotide or peptide sequence. The term 'Primary structure' was first coined by Linderstrøm-Lang in his 1951 Lane Medical Lectures.

Amino acid	In chemistry, an Amino acid is a molecule containing both amine and carboxyl functional groups. These molecules are particularly important in biochemistry, where this term refers to alpha-amino acids with the general formula $H_2NCHRCOOH$, where R is an organic substituent. In the alpha amino acids, the amino and carboxylate groups are attached to the same carbon, which is called the α-carbon.
Peptide	Peptides are short polymers formed from the linking, in a defined order, of α-amino acids. The link between one amino acid residue and the next is known as an amide bond or a Peptide bond. Proteins are polypeptide molecules .
Stereochemistry	Stereochemistry, a subdiscipline of chemistry, involves the study of the relative spatial arrangement of atoms within molecules. An important branch of Stereochemistry is the study of chiral molecules . Stereochemistry is a hugely important facet of chemistry and the study of stereochemical problems spans the entire range of organic, inorganic, biological, physical and supramolecular chemistries.
Complete protein	A Complete protein (or whole protein) is a source of protein that contains an adequate proportion of all of the essential amino acids for the dietary needs of humans or other animals. This does not refer to the protein source only containing all the essential amino acids, but also containing them in complete proportion for use by the human body. A source may contain all essential amino acids, but contain one in lower proportion to the others, making it an in Complete protein .
Essential amino acid	An Essential amino acid or indispensable amino acid is an amino acid that cannot be synthesized de novo by the organism (usually referring to humans), and therefore must be supplied in the diet. Eight amino acids are generally regarded as essential for humans: phenylalanine, valine, threonine, tryptophan, isoleucine, methionine, leucine, and lysine. Cysteine (or sulphur-containing amino acids), tyrosine (or aromatic amino acids), histidine and arginine are additionally required by infants and growing children.
Glycin	Glycin is N-substituted p-aminophenol. It is a photographic developing agent used in classic B'W developer solutions. It is a derivative of the amino acid Glycin e.
Glycine	Glycine is the organic compound with the formula NH_2CH_2COOH. It is the smallest of the 20 amino acids commonly found in proteins, coded by codons GGU, GGC, GGA and GGG. Glycine is unique among the proteinogenic amino acids in that it is not chiral. Most proteins incorporate only small quantities of Glycine. A notable exception is collagen, which contains about 35% Glycine.
Petroleum	Petroleum or crude oil is a naturally occurring, flammable liquid found in rock formations in the Earth consisting of a complex mixture of hydrocarbons of various molecular weights, plus other organic compounds. The term 'Petroleum' was first used in the treatise De Natura Fossilium, published in 1546 by the German mineralogist Georg Bauer, also known as Georgius Agricola.

The proportion of hydrocarbons in the mixture is highly variable and ranges from as much as 97% by weight in the lighter oils to as little as 50% in the heavier oils and bitumens.

Amination

Amination is the process by which an amine group is introduced into an organic molecule. This can occur in a number of ways including reaction with ammonia or another amine such as an alkylation, reductive Amination and the Mannich reaction. For example, -COOH --> $-CONH_2$.

Reductive amination

Reductive amination is a chemical reaction which involves the conversion of a carbonyl group to an amine, via an intermediate imine. The carbonyl group is most commonly a ketone or an aldehyde.

In this organic reaction, the amine first reacts with the carbonyl group to form a hemiaminal species, which subsequently loses one molecule of water in a reversible manner by alkylimino-de-oxo-bisubstitution, to form the imine.

Transamination

There are two chemical reactions known as Transamination The first is the reaction between an amino acid and an alpha-keto acid. The amino group is transferred from the former to the latter; this results in the amino acid being converted to the corresponding α-keto acid, while the reactant α-keto acid is converted to the corresponding amino acid (if the amino group is removed from an amino acid, an α-keto acid is left behind.)

Alpha carbon

The alpha carbon in organic chemistry refers to the first carbon that attaches to a functional group (the carbon is attached at the first position.) By extension, the second carbon is the beta carbon, and so on. This nomenclature can also be applied to the hydrogen atoms attached to the carbons.

Alpha cleavage

Alpha cleavage, (α-cleavage) in organic chemistry, refers to the act of breaking the carbon-carbon bond, adjacent to the carbon bearing a specified functional group.

Generally this topic is discussed when covering mass spectrometry and occurs generally by the same mechanisms.

For example of a mechanism of Alpha cleavage, an electron is knocked off an atom (usually by electron collision) to form a radical cation.

Esters

Esters are chemical compounds derived formally from an oxoacid (one containing an oxo group, X=O), and a hydroxyl compound such as an alcohol or phenol. Esters consist of an inorganic acid or organic acid in which at least one -OH (hydroxyl) group is replaced by an -O-alkyl (alkoxy) group. They are analogous to salts, using organic alcohols instead of metallic hydroxides.

Acylation

In chemistry, Acylation is the process of adding an acyl group to a compound. The compound providing the acyl group is called the acylating agent.

Because they form a strong electrophile when treated with some metal catalysts, acyl halides are commonly used as acylating agents.

Amide

In chemistry, an Amide is one of three kinds of compounds:

- (sometimes called acid Amide the organic functional group characterized by a carbonyl group (C=O) linked to a nitrogen atom (N), or a compound that contains this functional group (pictured to the right); or
- a kind of anion or
- any organic compound derived by the replacement of a hydroxyl group by an amino group.

Amide s are the most stable of all the carbonyl functional groups.

Many chemists make a pronunciation distinction between the two, saying /É™ËˆmiË□d/ for the carbonyl-nitrogen compound and /ËˆæmaÉªd/ for the anion. Others substitute one of these with , while still others pronounce both , making them homonyms.

In the first sense referred to above, an Amide is an amine where one of the nitrogen substituents is an acyl group; it is generally represented by the formula: $R_1(CO)NR_2R_3$, where either or both R_2 and R_3 may be hydrogen.

Carboxyl group

A Carboxyl group is a set of four atoms bonded together and present in carboxylic acids, including amino acids. Usually abbreviated as either CO_2H or COOH, this set of atoms constitutes a functional group. In every Carboxyl group the carbon atom is attached to an oxygen atom by a double bond and to a hydroxyl group (OH) by a single bond.

Histamine

Histamine is a biogenic amine involved in local immune responses as well as regulating physiological function in the gut and acting as a neurotransmitter. Histamine triggers the inflammatory response. As part of an immune response to foreign pathogens, Histamine is produced by basophils and by mast cells found in nearby connective tissues.

Ninhydrin

Ninhydrin is a chemical used to detect ammonia or primary and secondary amines. When reacting with these free amines, a deep blue or purple color known as Ruhemann's purple is evolved. Ninhydrin is most commonly used to detect fingerprints, as amines left over from peptides and proteins (terminal amines or lysine residues) sloughed off in fingerprints react with Ninhydrin.

Dipeptide

A Dipeptide is a molecule consisting of two amino acids joined by a single peptide bond.

Dipeptide s are produced from polypeptides by the action of the hydrolase enzyme dipeptidyl peptidase. Dietary proteins are digested to Dipeptide s and amino acids, and the Dipeptide s are absorbed more rapidly than the amino acids, because their uptake involves a separate mechanism.

Bradykinin

Bradykinin is a nonapeptide that causes blood vessels to enlarge (dilate), and therefore causes blood pressure to lower. A class of drugs called ACE inhibitors, which are used to lower blood pressure, increase Bradykinin further lowering blood pressure. Bradykinin works on blood vessels through the release of prostacyclin, nitric oxide, and Endothelium-Derived Hyperpolarizing Factor.

Cystine

Cystine is the amino acid dimer formed when a pair of cysteine molecules are joined by a disulfide bond. It is described by the formula $(SCH_2CH(NH_2)CO_2H)_2$. It is a colorless solid, and melts at 247-249 °C. It was discovered in 1810 by William Hyde Wollaston but was not recognized as a component of proteins until it was isolated from the horn of a cow in 1899.

Oxytocin

Oxytocin is a mammalian hormone that also acts as a neurotransmitter in the brain.

It is best known for its roles in female reproduction: it is released in large amounts after distension of the cervix and vagina during labor, and after stimulation of the nipples, facilitating birth and breastfeeding, respectively. Recent studies have begun to investigate Oxytocin's role in various behaviors, including orgasm, social recognition, pair bonding, anxiety, trust, love, and maternal behaviors.

Orexin

Orexins are the common names given to a pair of excitatory neuropeptide hormones that were simultaneously discovered by two groups of researchers in rat brains.

The two related peptides, with approximately 50% sequence identity, are produced by cleavage of a single precursor protein. Orexin-A/hypocretin-1 is 33 amino acid residues long and has two intrachain disulfide bonds, while Orexin-B/hypocretin-2 is a linear 28 amino acid residue peptide.

Insulin

Insulin is a hormone that has extensive effects on metabolism and other body functions, such as vascular compliance. Insulin causes cells in the liver, muscle, and fat tissue to take up glucose from the blood, storing it as glycogen in the liver and muscle, and stopping use of fat as an energy source. When Insulin is absent (or low), glucose is not taken up by body cells, and the body begins to use fat as an energy source, for example, by transfer of lipids from adipose tissue to the liver for mobilization as an energy source.

Eclipsed conformation

In chemistry an Eclipsed conformation is a chemical conformation that exists in any open chain single chemical bond connecting two sp^3 hybridised atoms as a conformational energy maximum. This maximum is often explained by steric hindrance, but its origins sometimes actually lie in hyperconjugation (as when the eclipsing interaction is of two hydrogen atoms.) Staggered conformation image right in Newman projection Eclipsed conformation

In the example of ethane in Newman projection it shows that rotation around the carbon-carbon bond is not entirely free but that an energy barrier exists.

Carboxypeptidase

Carboxypeptidase (EC number 3.4.16 - 3.4.18) is an enzyme that hydrolyzes the carboxy-terminal (C-terminal) end of a peptide bond. Humans, animals, and plants contain several types of Carboxypeptidase s with diverse functions ranging from catabolism to protein maturation.

The first Carboxypeptidase s studied were those involved in the digestion of food (pancreatic Carboxypeptidase s A1, A2, and B.)

Enzymes

Enzymes are biomolecules that catalyze (i.e., increase the rates of) chemical reactions. Nearly all known Enzymes are proteins. However, certain RNA molecules can be effective biocatalysts too.

Peptide synthesis

In organic chemistry, Peptide synthesis is the production of peptides, which are organic compounds in which multiple amino acids are linked via peptide bonds which are also known as amide bonds. The biological process of producing long peptides (proteins) is known as protein biosynthesis.

Peptides are synthesized by coupling the carboxyl group or C-terminus of one amino acid to the amino group or N-terminus of another.

Conjugated protein

A Conjugated protein is a protein that functions in interaction with other chemical groups attached by covalent bonds or by weak interactions.

Many proteins contain only amino acids and no other chemical groups, and they are called simple proteins. However, other kind of proteins yield, on hydrolysis, some other chemical component in addition to amino acids and they are called Conjugated protein s.

Primary structure

In biochemistry, the Primary structure of a biological molecule is the exact specification of its atomic composition and the chemical bonds connecting those atoms (including stereochemistry.) For a typical unbranched, un-crosslinked biopolymer (such as a molecule of DNA, RNA or typical intracellular protein), the Primary structure is equivalent to specifying the sequence of its monomeric subunits, e.g., the nucleotide or peptide sequence. The term 'Primary structure' was first coined by Linderstrøm-Lang in his 1951 Lane Medical Lectures.

Secondary structure

In biochemistry and structural biology, Secondary structure is the general three-dimensional form of local segments of biopolymers such as proteins and nucleic acids (DNA/RNA.) It does not, however, describe specific atomic positions in three-dimensional space, which are considered to be tertiary structure.

Secondary structure is formally defined by the hydrogen bonds of the biopolymer, as observed in an atomic-resolution structure.

Radicals

In chemistry, Radicals are atoms, molecules, or ions with unpaired electrons on an otherwise open shell configuration. These unpaired electrons are usually highly reactive, so Radicals are likely to take part in chemical reactions. Radicals play an important role in combustion, atmospheric chemistry, polymerization, plasma chemistry, biochemistry, and many other chemical processes, including human physiology.

Tertiary structure

In biochemistry and chemistry, the Tertiary structure of a protein or any other macromolecule is its three-dimensional structure, as defined by the atomic coordinates.

Tertiary structure is considered to be largely determined by the protein's primary structure, or the sequence of amino acids of which it is composed. Efforts to predict Tertiary structure from the primary structure are known generally as protein structure prediction.

Quaternary structure

In biochemistry, Quaternary structure is the arrangement of multiple folded protein molecules in a multi-subunit complex.

Many proteins are actually assemblies of more than one polypeptide chain, which in the context of the larger assemblage are known as protein subunits. In addition to the tertiary structure of the subunits, multiple-subunit proteins possess a Quaternary structure, which is the arrangement into which the subunits assemble.

Chaperone

In molecular biology, chaperones are proteins that assist the non-covalent folding/unfolding and the assembly/disassembly of other macromolecular structures, but do not occur in these structures when the latter are performing their normal biological functions. The common perception that chaperones are primarily concerned with protein folding is incorrect. The first protein to be called a Chaperone assists the assembly of nucleosomes from folded histones and DNA and such assembly chaperones, especially in the nucleus, are concerned with the assembly of folded subunits into oligomeric structures.

Lipids

Lipids are a broad group of naturally-occurring molecules which includes fats, waxes, sterols, fat-soluble vitamins (such as vitamins A, D, E and K), monoglycerides, diglycerides, phospholipids, and others. The main biological functions of Lipids include energy storage, as structural components of cell membranes, and as important signaling molecules.

Lipids may be broadly defined as hydrophobic or amphiphilic small molecules; the amphiphilic nature of some Lipids allows them to form structures such as vesicles, liposomes, or membranes in an aqueous environment.

Simple lipid

A Simple lipid is a saponifiable lipid with just two types of components. When more components exist, the saponifiable lipid is classified as a complex lipid.

Wax

Wax refers to beeswax or another substance with similar properties. The traditional meaning, beeswax, refers to a substance secreted by bees and used by them in constructing their honeycombs. The term has come to refer more generally to a class of substances with properties similar to beeswax, in respect of

- plastic (malleable) at normal ambient temperatures
- a melting point above approximately 45 °C (113 °F) (which differentiates waxes from fats and oils)
- a relatively low viscosity when melted (unlike many plastics)
- insoluble in water
- hydrophobic

Waxes may be natural secretions of plants or animals, artificially produced by purification from natural petroleum or completely synthetic. In addition to beeswax, carnauba (a plant epicuticular Wax) and paraffin (a petroleum Wax) are commonly encountered waxes which occur naturally.

CD36

CD36 is an integral membrane protein found on the surface of many cell types in vertebrate animals and is also known as FAT, SCARB3, GP88, glycoprotein IV and glycoprotein IIIb CD36 is a member of the class B scavenger receptor family of cell surface proteins. CD36 binds many ligands including collagen, thrombospondin, erythrocytes parasitized with Plasmodium falciparum, oxidized low density lipoprotein, native lipoproteins, oxidized phospholipids, and long-chain fatty acids.

Fatty acid

In chemistry, especially biochemistry, a Fatty acid is a carboxylic acid often with a long unbranched aliphatic tail (chain), which is either saturated or unsaturated. Carboxylic acids as short as butyric acid (4 carbon atoms) are considered to be Fatty acid s, whereas Fatty acid s derived from natural fats and oils may be assumed to have at least eight carbon atoms, caprylic acid (octanoic acid), for example. The most abundant natural Fatty acid s have an even number of carbon atoms because their biosynthesis involves acetyl-CoA, a coenzyme carrying a two-carbon-atom group

Fatty acids

Fatty acids are an important source of energy for many organisms. Excess glucose can be stored efficiently as fat. Triglycerides yield more than twice as much energy for the same mass as do carbohydrates or proteins.

Glycerides

Glycerides more correctly known as acylglycerols, are esters formed from glycerol and fatty acids.

Glycerol has three hydroxyl functional groups, which can be esterified with one, two, or three fatty acids to form mono Glycerides , di Glycerides , and tri Glycerides .

	Vegetable oils and animal fats contain mostly tri Glycerides , but are broken down by natural enzymes (lipases) into mono- and di Glycerides and free fatty acids.
Triglycerides	Â·) (more properly known as Â·), TAG or triacylglyceride) is a glyceride in which the glycerol is esterified with three fatty acids. It is the main constituent of vegetable oil and animal fats. Triglycerides are formed from a single molecule of glycerol, combined with three fatty acids on each of the OH groups, and make up most of fats digested by humans.
Lecithin	Lecithin is any of a group of yellow-brownish fatty substances occurring in animal and plant tissues, and in egg yolk, composed of phosphoric acid, choline, fatty acids, glycerol, glycolipids, triglycerides, and phospholipids (e.g., phosphatidylcholine, phosphatidylethanolamine, and phosphatidylinositol.) However, Lecithin is sometimes used as a synonym for pure phosphatidylcholine, a phospholipid that is the major component of its phosphatide fraction. It may be isolated either from egg yolk or from soy beans, from which it is extracted chemically or mechanically.
Phosphatidic acid	Phosphatidic acid is the acid form of phosphatidate, a common phospholipid that is a major constituent of cell membranes. Phosphatidate acid is the smallest of the phospholipids. Phosphatidic acid consists of a glycerol backbone, with, in general, a saturated fatty acid bonded to carbon-1, an unsaturated fatty acid bonded to carbon-2, and a phosphate group bonded to carbon-3.
Phospholipids	Phospholipids are a class of lipids and are a major component of all cell membranes. Most Phospholipids contain a diglyceride, a phosphate group, and a simple organic molecule such as choline; one exception to this rule is sphingomyelin, which is derived from sphingosine instead of glycerol. They are a type of molecule.
Choline	Choline is an organic compound, classified as a water-soluble essential nutrient and usually grouped within the Vitamin B complex. This natural amine is found in the lipids that make up cell membranes and in the neurotransmitter acetyl Choline . Adequate intakes (AI) for this micronutrient of between 425 to 550 milligrams daily, for adults, have been established by the Food and Nutrition Board of the Institute of Medicine of the National Academy of Sciences.
Steroid	A Steroid is a terpenoid lipid characterized by a carbon skeleton with four fused rings, generally arranged in a 6-6-6-5 fashion. Steroids vary by the functional groups attached to these rings and the oxidation state of the rings. Hundreds of distinct steroids are found in plants, animals, and fungi.
Androsterone	Androsterone is a steroid hormone with weak androgenic activity. It is made in the liver from the metabolism of testosterone. It was first isolated in 1931, by Adolf Friedrich Johann Butenandt and Kurt Tscherning.
Cholesterol	Cholesterol is a lipidic, waxy steroid found in the cell membranes and transported in the blood plasma of all animals. It is an essential component of mammalian cell membranes where it is required to establish proper membrane permeability and fluidity. Cholesterol is the principal sterol synthesized by animals, but small quantities are synthesized in other eukaryotes, such as plants and fungi.

Hormone	Hormone s are chemicals released by cells that affect cells in other parts of the body. Only a small amount of Hormone is required to alter cell metabolism. It is essentially a chemical messenger that transports a signal from one cell to another.
Essential oil	An Essential oil is a concentrated, hydrophobic liquid containing volatile aroma compounds from plants. Essential oil s are also known as volatile or ethereal oils, or simply as the 'oil of' the plant from which they were extracted, such as oil of clove. An oil is 'essential' in the sense that it carries a distinctive scent, or essence, of the plant.
Terpene	Terpenes are a large and varied class of hydrocarbons, produced primarily by a wide variety of plants, particularly conifers, though also by some insects such as termites or swallowtail butterflies, which emit terpenes from their osmeterium. They are the major components of resin, and of turpentine produced from resin. The name 'Terpene' is derived from the word 'turpentine'.
Carotene	The term Carotene is used for several related substances having the formula $C_{40}H_x$, which are synthesized by plants but cannot be made by animals. Carotene is an orange photosynthetic pigment important for photosynthesis. Carotene s are responsible for the orange colour of the carrot for which it is named, and many other fruits and vegetables (for example, sweet potatoes and orange cantaloupe melon.)
Monoterpene	Monoterpenes are a class of terpenes that consist of two isoprene units and have the molecular formula $C_{10}H_{16}$. Monoterpenes may be linear (acyclic) or contain rings. Biochemical modifications such as oxidation or rearrangement produce the related monoterpenoids.
Myrcene	Myrcene is an olefinic natural organic compound. It is classified as a hydrocarbon and a monoterpene. It is obtained from the essential oil of the plants bay, verbena, myrcia (from which is gets its name) and others.
Sesquiterpenes	Sesquiterpenes are a class of terpenes that consist of three isoprene units and have the molecular formula $C_{15}H_{24}$. Like monoterpenes, Sesquiterpenes may be acyclic or contain rings, including many unique combinations. Biochemical modifications such as oxidation or rearrangement produce the related sesquiterpenoids.
Retinol	Retinol, the animal form of vitamin A, is a fat-soluble vitamin important in vision and bone growth. It is also a diterpenoid. Retinol is among the most usable forms of vitamin A, which also include Retinal (aldehyde form), Retinoic acid (acid form) and retinyl ester (ester form.)
Terpenoids	The Terpenoids , sometimes called isoprenoids, are a large and diverse class of naturally-occurring organic chemicals similar to terpenes, derived from five-carbon isoprene units assembled and modified in thousands of ways. Most are multicyclic structures that differ from one another not only in functional groups but also in their basic carbon skeletons. These lipids can be found in all classes of living things, and are the largest group of natural products.

Halogenation	Halogenation is a chemical reaction that incorporates a halogen atom into a molecule. More specific descriptions exist that specify the type of halogen: fluorination, chlorination, bromination, and iodination. In a Markovnikov addition reaction, a halogen like bromine is reacted with an alkene which causes the π-bond to break forming an haloalkane.
Stereochemistry	Stereochemistry, a subdiscipline of chemistry, involves the study of the relative spatial arrangement of atoms within molecules. An important branch of Stereochemistry is the study of chiral molecules . Stereochemistry is a hugely important facet of chemistry and the study of stereochemical problems spans the entire range of organic, inorganic, biological, physical and supramolecular chemistries.
Nylon	Nylon is a generic designation for a family of synthetic polymers known generically as polyamides and first produced on February 28, 1935 by Wallace Carothers at DuPont. Nylon is one of the most commonly used polymers. Nylon is a thermoplastic silky material, first used commercially in a Nylon-bristled toothbrush (1938), followed more famously by women's stockings ('nylons'; 1940.)
Polyamide	A Polyamide is a polymer containing monomers of amides joined by peptide bonds. They can occur both naturally, examples being proteins, such as wool and silk, and can be made artificially through step-growth polymerization, examples being nylons, aramids, and sodium poly(aspartate.) The amide link is produced from the condensation reaction of an amino group and a carboxylic acid or acid chloride group.
Carbonate ester	A Carbonate ester is a functional group in organic chemistry consisting of a carbonyl group flanked by two alkoxy groups. The general structure of these carbonates is $R_1O(C=O)OR_2$ and they are related to esters $R_1O(C=O)R$ and ethers R_1OR_2 and also to the inorganic carbonates. Carbonate ester s are used as protecting group for diols.
Polycarbonates	Polycarbonates are a particular group of thermoplastic polymers. They are easily worked, moulded, and thermoformed; as such, these plastics are very widely used in the modern chemical industry. Their interesting features (temperature resistance, impact resistance and optical properties) position them between commodity plastics and engineering plastics.
Polyester	Polyester is a category of polymers which contain the ester functional group in their main chain. Although there are many polyesters, the term 'Polyester' as a specific material most commonly refers to polyethylene terephthalate (PET.) Polyesters include naturally-occurring chemicals, such as in the cutin of plant cuticles, as well as synthetics through step-growth polymerization such as polycarbonate and polybutyrate.
Polyurethane	A Polyurethane, commonly abbreviated PU, is any polymer consisting of a chain of organic units joined by urethane (carbamate) links. Polyurethane polymers are formed through step-growth polymerization by reacting a monomer containing at least two isocyanate functional groups with another monomer containing at least two hydroxyl (alcohol) groups in the presence of a catalyst.

	Polyurethane formulations cover an extremely wide range of stiffness, hardness, and densities.
Ethyl carbamate	Ethyl carbamate is a substance first prepared in the nineteenth century. Structurally it is an ester of carbamic acid, i. e., Ethyl carbamate as shown.

CPSIA information can be obtained
at www.ICGtesting.com
Printed in the USA
LVOW03s1317080316
478273LV00008B/250/P

9 781428 881716